W9-BPS-724

A SCIENTIST'S
GUIDE TO
TALKING WITH
THE MEDIA

A SCIENTIST'S GUIDE TO TALKING WITH THE MEDIA

Practical Advice from the Union of Concerned Scientists

RICHARD HAYES AND
DANIEL GROSSMAN

RUTGERS UNIVERSITY PRESS
NEW BRUNSWICK, NEW JERSEY, AND LONDON

LIBRARY OF CONGRESS CATALOGING-IN-PUBLICATION DATA

Hayes, Richard, 1968–

A scientist's guide to talking with the media : practical advice from
the Union of Concerned Scientists / Richard Hayes and Daniel Grossman.

p.   cm.

Includes bibliographical references and index.

ISBN-13: 978-0-8135-3857-0 (alk. paper)

ISBN-13: 978-0-8135-3858-7 (pbk. : alk. paper)

1. Science in mass media. 2. Science news. 3. Science—Public opinion.
4. Communication in science. I. Grossman, Daniel, 1959–
II. Union of Concerned Scientists. III. Title.

Q225.H39 2006   500—dc22   2005035567

A British Cataloging-in-Publication record for this book
is available from the British Library

Book design by Kevin Hanek
Composition by Kevin Hanek and Kathy Geary
Set in the OpenType version of Adobe Kepler

Manufactured in the United States of America

Contents

Acknowledgments

T HIS BOOK WOULD not be possible without the support of the leadership and staff of the Union of Concerned Scientists. We are especially grateful to Suzanne Shaw for all her hard work on our behalf, and Kevin Knobloch for providing the resources to make it happen. Katherine Moxhet made our lives much easier with her tireless research efforts and constant good humor. Philip Knowles was a valuable research assistant as well. Bryan Wadsworth and Heather Tuttle went beyond the call of duty by editing and organizing portions of the book even when other work beckoned. Thanks also to Eileen Quinn, Paul Fain, Eric Young, Luke Warren, and Linda Gunter for all their insights into the world of science media relations. We are indebted to Jane Rissler for reviewing portions of the manuscript from a scientist's perspective, and to Deborah Blum, for her helpful comments.

To our agent, Faith Hamlin, thank you for securing Rutgers University Press as our partner on this book. We consider ourselves lucky to have an editor as thoughtful and meticulous as Audra Wolfe. Thanks also to Anne Schneider, who copyedited the final manuscript, and Beth Kressel and Nicole Manganaro, who shepherded the book through the final hoops.

We are continually grateful for the support and encouragement of our wives, Meghan Hayes and Sarah Bansen, and our families.

Finally, we owe a galactic-sized thank you to all the scientists, journalists, and communication experts who shared their expertise and experiences with us. This book is yours as much as ours.

Introduction

HERE'S A POP QUIZ. True or false:

- Dinosaurs and humans never lived side by side.
- Molecules are bigger than electrons.
- Lasers work by focusing light waves.
- Earth goes around the sun once a year.
- Antibiotics can't kill viruses.

Scientists may know that every one of these statements is true, but unfortunately that can't be said of Americans as a whole. In fact, for each statement, about half of the adults polled by the National Science Foundation (NSF) in 2001 gave the wrong answer.[1] The federal agency's biennial report on public attitudes and understanding of science and technology found not only that the U.S. public is misinformed about basic scientific facts, but also that around two-thirds of Americans don't really understand how science works.[2]

It makes sense, accordingly, that a hefty fraction of Americans accept pseudo-scientific claims—assertions presented in scientific language but lacking evidence and plausibility—about the existence of ghosts and telepathic communication and the like.[3] For instance, more than a quarter of Americans believe the premise of astrology: that the positions of stars and planets influence a person's life.[4] Belief in such "paranormal" phenomena is on the rise, according to the NSF.

Other studies reinforce the notion that Americans are ignorant of basic scientific and technological principles. For example, only 12 percent of Americans surveyed by the National Environmental Education & Training Foundation and Roper

Research passed a basic quiz about U.S. energy production and use.[5] Barely half of Americans surveyed by the International Technology Education Association (ITEA) knew how a telephone gets sound from one location to another.

Perhaps this should not come as a surprise, since as a species we have always let specialists attend to the details of technology so that the rest of us don't have to rub sticks together to light fires. Even back in 1936, Albert Einstein said, "Just ask a hundred people off the street what really happens in a telephone apparatus or in a radio receiver and how the electric current is produced which lights their room at night. Then you will see that most of them live like strangers in the world of the things entrusted to them."[6] Or as the National Research Council concluded in a report on this topic, "As a society, we are not even fully aware of or conversant with the technologies we use every day. In short, we are not 'technologically literate.'"[7]

Even if this is not news to scientists, it's a sobering reality for those scientists who want to communicate their research or views to the general public. It also has implications if public policies are to be built on a firm empirical foundation: can there be an informed debate about the threat of or protection against biological warfare, for instance, if the most basic facts about treatment against infection are misunderstood? Can there be a consensus on how to stop global warming when a sizable fraction of the electorate doesn't understand the relationship in space between Earth and the sun?

The purpose of this book is not to help to educate the American public on the principles of scientific thought, the bedrock facts of scientific understanding, or the basic principles of technology. That is a long-term goal that will require a dedicated effort by many parties. But does the public really need to know how everything works to appreciate the value of scientific discovery and related activities? Certainly not. Americans understand the value of telephony, for example, without understanding exactly what's inside a telephone. We believe that the public wants to know what the latest advances in science and technology mean for people, businesses, or the natural world even when they don't understand how the science or technology works. Scientists and other researchers, therefore, can play an important role not only in communicating the facts, but also the *value* of science and technology.

This book focuses specifically on how scientists can use the media to raise awareness of scientific research and thought. Most adults in the United States learn about newly proposed theories, newly discovered facts, and established knowledge through the media. By one estimate, kids get 83 percent of their environmental information through the media.[8] Even scientists learn about scientific advances—including in their own fields—from newspapers, radio, and television (and, to an increasing degree, Internet news sites). Lee Frelich, an ecologist at the Twin Cities campus of the University of Minnesota, encapsulated why the media are so important when he said of his outreach to print and broadcast journalists, "I can reach a larger number of people than I will at classes at the university or at science meetings." Frelich estimates that he has been interviewed sixty or seventy times on the ecological impact of invasive earthworms, on the differences between virgin and second-growth forests, and on other ecological issues for *National Geographic*, the *Wall Street Journal*, CBS News, and dozens of other national and local newspapers, magazines, and television shows.

The media are a sort of information utility serving up a small but steady stream of science news, and the scientists who appear in quotes or sound bites or behind the scenes as crucial sources of information are the fuel. But what if the fuel line is crimped? Like electric utilities during a power failure, this public service would shut down and the public would be left in the dark. That is in part what is happening today as changing priorities in the news business have reduced science coverage and budgets. Scientists must redouble their efforts to be heard, but so far they have not. The late mathematician and humanist Jacob Bronowski would have deplored such a state of affairs. He considered science literacy to be a matter of urgent necessity, writing that "The world today is made, it is powered by science; and for any [person] to abdicate an interest in science is to walk with open eyes toward slavery."[9]

Part of the problem is that many scientists are not comfortable in the role of communicator. Many scientists are reluctant even to talk to reporters. In chapter 2, we discuss in detail the reasons why. The most obvious is prudence. Every time a scientist appears in the media, he or she is putting a painstakingly acquired reputation on the line. A scientific reputation is built slowly and carefully over years. Bad newspaper articles or television appearances can tarnish that reputation. Prudence, among other things, warns scientists to keep out of the limelight. But you can't be a shrinking violet; your research and knowledge are too important to be consumed solely by your colleagues. Each interview, each press release, each photo opportunity is an act of adult education that gradually—like grains of windblown sand that imperceptibly turn into a sand dune—builds a foundation of principles and facts to support scientific understanding. Scientists must speak out.

And that's where *A Scientist's Guide to Talking with the Media* comes in. This book is intended for you, the working scientist, which we take to include social scientists, hard scientists, medical professionals, and engineers, working in academia, the private sector, nonprofit institutions, and the government. We hope others (including journalism and science students, academics of all stripes, public information officers, and reporters) find it useful as well. We surveyed hundreds of scientists across the country to uncover what works when speaking to the media and what doesn't. We conducted in-depth interviews with scores of scientists and dozens of reporters. We couldn't include all the stories they shared with us, so we sought to highlight those that were either representative of the experiences of others or best illustrated an important point. It is these personal experiences and words of advice—in addition to those of the authors, science media relations expert Richard Hayes and science journalist Daniel Grossman—that shape the contents of this book.

Our main goal is to give you the tools and information you need to make the best media decisions possible—decisions that will not only increase the likelihood of media coverage for your work or views, but also increase the chances that the coverage is factually and contextually accurate. The advice we offer is tailored to the needs and constraints of both scientists and journalists. We have discovered, for example, that some of the most common roadblocks to communicating effectively with reporters arise from simple misunderstandings about how they do their jobs. The media comprise very different sectors—newspapers, magazines, television, radio, the Internet, etc.—each with its own idiosyncrasies. With a better understanding of how news stories are produced, and of the pressures on those gathering the news, scientists will be more likely to get their points of view across in a way that provides reporters what they need.

Effective communication increases the odds that today's story will be told well, and can also make you a prized source for the harried reporter who'll want to call on you in the future. In chapter 3, we lend a helping hand by explaining how reporters approach scientific issues and how their jobs have become harder than ever due to a variety of forces within their organizations and on the street.

The advice we offer may well require that you work with the media differently than you have in the past (if you worked with reporters at all). But we want to be clear: we are not suggesting that you transform yourself into anything you are not. The task of getting more and better coverage for your work and for science in general is not about remaking you in the image of those who most dominate the media: politicians, celebrities, and athletes. Despite the often-unflattering stereotypes that lump scientists into one homogeneous group, we realize you are a diverse lot. Some scientists are outgoing and gregarious while others are introverted and reluctant public speakers. The rest fall somewhere in between. Whether or not you are comfortable speaking about your work with non-scientists, our advice is premised on the fundamental principles that make you a scientist: intellectual honesty, forthrightness, and rigor.

To communicate more effectively, you must present your research in ways the media can use and the public can understand. In chapter 4, we explain why scientists must boil down their messages to a few central points to ensure reporters don't make mistakes. It is possible to simplify without becoming simplistic, and this chapter shows you how to do it by turning themes and talking points into sound bites.

In chapter 5, we explain how to take control of the interview and stay "on message." Some of our advice applies to all media. In other cases we discuss techniques that apply exclusively to each of the major media in turn, including radio, television, and print. We provide tips from journalists, public relations experts, and other scientists, we illustrate our points with case studies, and we highlight common mistakes. Getting down to the nitty-gritty, we even give you practical style pointers so you'll know what to wear when you're interviewed on television.

Teenagers know that before you date you have to know how to flirt. By the same token, before you can talk to reporters you have to know how to get their attention. In chapter 6, we show you how to become a trusted source for reporters, and in chapter 7 we introduce you to the tools you need to interest the media in you and your work and how to use these tools effectively. We explain how to write a press release, how to construct an op-ed that stands a good chance of getting printed in the newspaper, why a well-written letter to the editor is probably more powerful than you think, and several other ways you can get your research and views out to the public through the media.

Finally, in chapter 8, we address the pros and cons of being more outspoken in the media, especially related to matters of broad public concern such as U.S. government policy. The number of scientists in the United States today who serve as unofficial public science consultants or commentators, speaking on a wide range of technical topics in high-profile hearings and the national news, is small indeed. The late Carl Sagan, that turtleneck-wearing ambassador of astronomy and space exploration, was one of the country's best-known scientists for a generation. He not only popularized and promoted astronomy and the search for extraterrestrial intelligence, but also became an advocate for science itself. Many in the scientific community frowned on his extracurricular activities in the media, but Sagan replied by inviting more scientists to join him: "Having only a few visible scientists, if that's the way it works out, is not good, but it's better than having no visible scientists." We highlight the stories of a few highly visible scientists working today, showing both the difficulties they face and the satisfaction they've achieved.

First, though, we consider in chapter 1 why scientists should talk to the media in the first place.

A SCIENTIST'S
GUIDE TO
TALKING WITH
THE MEDIA

CHAPTER 1

We Need to Talk

INDULGE US FOR just a moment with a gender stereotype and imagine a husband and wife trying to salvage a troubled marriage. Alone in a room, they're going over the same bitter ground they've covered again and again each time the subject of their relationship comes up.

> He: "You're just not accessible."
> She: "I try but you don't seem to understand me."
> He: "If only you would talk to me in plain English."
> She: "You just want me to oversimplify."

Now, in your mind's eye, exchange "He" with a journalist and "She" with a scientist. The notion that people with science training and people with literary training speak in different tongues was crystallized in 1959 in the speech, "The Two Cultures and the Scientific Revolution," by the late scientist and novelist C. P. Snow. Snow's controversial contention that an intellectual fault line separates scientists from humanists has been debated endlessly. Even if this were true, it could not possibly apply to every member of the postulated cultures (as Snow's own resume testifies), but there is enough supporting evidence—some coming from solid research—to make Snow's contention worth considering. The insight gained could help improve relations between the sometimes estranged, if not completely distinct, cultures of science and journalism.

A SIMPLE MISUNDERSTANDING

Doctors surveyed in 2003 by Dr. Teresa Schrader, then an editorial fellow at the New England Journal of Medicine, were unimpressed by the quality of medical information in the news. Eighty percent of 408 doctors who responded to the survey said that health reporting was no better than "fair," and even more said that health news "misleads or confuses patients or disrupts their decision-making." In a 2004 email poll of scientists conducted by the Union of Concerned Scientists, 90 percent of respondents said the media does a poor job covering science. More than half reported having difficult or disappointing experiences with the press.[1]

These findings echo the results of a 1997 study by Vanderbilt University's First Amendment Center on relations between scientists and journalists.[2] The authors, Jim Hartz (a veteran science journalist) and Rick Chappell (a NASA scientist), were so vexed by the strains between the two professions that they titled their report *Worlds Apart*. Based in part on an opinion survey conducted for the study, the authors reported a "gulf" in communication between "scientists who don't speak English and journalists who don't speak science."[3]

For instance, 85 percent of the journalists polled for the report considered scientists only "somewhat" or "not at all" accessible. Sixty-two percent of the journalists polled agreed with the strongly worded statement that scientists are "so intellectual and immersed in their own jargon that they can't communicate with journalists or the public." On the other side of the divide, more than 90 percent of the scientists polled believed that few members of the news media understand the nature of science and technology. Sixty-six percent said that most members of the press have no idea how to interpret scientific results, and 69 percent said most reporters have no understanding of scientific methods.[4] Hartz and Chappell go so far as to warn that the rift between those who create science and those who explain it to the public "threatens America's future."

Hartz and Chappell's findings are similar to the results of a study scientists conducted in Great Britain for the Wellcome Trust, the world's largest charitable funder of medical research.[5] Face-to-face interviews with 1,652 randomly selected scientists at university and public research institutes showed that only 6 percent of scientists would trust journalists working for national newspapers to "provide accurate information about scientific facts." The figure for television news was slightly higher, at 11 percent.[6]

Some science communication researchers say Hartz and Chappell exaggerate the tensions between scientists and reporters. First, 77 percent of the "journalists" polled for their report were actually editorial executives, not the reporters who meet scientists face-to-face in labs and who may have more positive views. Sharon Dunwoody, a journalism professor who was previously a science journalist, says Hartz and Chappell also neglected to consider whether the relationship between scientists and journalists is changing. Dunwoody has been studying these interactions since 1978 and has found that scientists are more knowledgeable about, and thus more tolerant of, the media today than when she began her research. "They are extremely savvy about these processes," she says. Lee Frelich, who was awarded his Ph.D. in biology in 1986, says that when he was a graduate student it was considered "unseemly" to appear in press accounts but "today that's changing." Deborah Blum, president of the National Association of Science Writers and a journalism professor at the University of Wisconsin-Madison, agrees. "Are we 'worlds apart' the way we once were?" Blum asks. "No."

Despite the methodological flaws and exaggerations in *Worlds Apart*, Blum, Dunwoody, and many other observers agree that a "clash of cultures" hinders the quality of science reporting and shortchanges public understanding (and support) of science. Mutual suspicion and wariness prevent scientists and journalists from developing better, trusting relationships. This "marriage," therefore, may be seriously strained, but divorce is not an option. The U.S. public relies on the combined efforts of scientists and journalists to learn about new findings in science and to reinforce old ones.

The problem, as a marriage counselor might say of a troubled union, could be attacked in part by efforts on both sides to understand—and accept—their differences. For instance, scientists who stop to consider the challenges of producing news on a deadline and getting up to speed on arcane topics in hours or at most days might be more sympathetic to the plight of reporters. Likewise, journalists who stop to consider the premium scientists place on numerical accuracy and the precise meaning of technical words (and the professional risks associated with imprecision of any sort) might be more understanding of interview subjects who insist on using jargon.

A number of paid fellowships aim to reduce misunderstandings between these two professions. For instance, the Knight Science Journalism Fellowship program at the Massachusetts Institute of Technology (MIT) offers "mid-career" journalists an academic year of seminars on the craft of science writing, field trips to laboratories, the chance to hobnob with scientists free of deadline pressure, and the ability to attend both introductory and advanced science classes. The Ted Scripps Fellowships in Environmental Journalism at the University of Colorado at Boulder (which one of the authors, Daniel Grossman, has received) is another such program.

Deborah Blum says the problem is less that journalists fail to appreciate the conventions of science but that scientists are in the dark about what it's like to be a journalist. Blum, who was a science writer at the *Sacramento Bee* (where she won a Pulitzer prize) for nearly two decades, says that while she and her colleagues are frequent visitors to research offices and labs, it is rare to see a scientist in the newsroom. "If I compared the amount of time I've spent with them versus how much they've spent with me it is off the charts," she says. Nobody says scientists and journalists should spend equal amounts of time in each other's work places. Nevertheless, there is a notable imbalance. The American Association for the Advancement of Science (AAAS) has made a modest attempt to redress it. Since 1975 the association has given science and engineering graduate students and postdoctoral researchers summer fellowships at major media outlets including National Public Radio, the *Los Angeles Times*, and *Scientific American*. The Aldo Leopold Leadership Program at Stanford University gives up to twenty scientists two weeks of media training every year. The AAAS and the American Geophysical Union, among other scientific societies, recognize members with annual awards for contributions to public awareness of science and its policy implications.

Scientific institutions in Great Britain have also been active in efforts to help scientists understand the media and improve their communication skills.[7] The Wellcome Trust study referenced earlier showed that British scientists believe the public has a relatively poor perception of them, which is no doubt one reason why British scientists feel that they need to improve their relations with the media. The study found that 44 percent of the scientists surveyed considered themselves "responsible," whereas only 9 percent thought the public considers them responsible—a "perception gap" of 35 percent. A similar gap of 32 percent was recorded between the number of scientists who considered their profession honest and their perception of what the public thinks.

The British Association for the Advancement of Science is trying to help scientists communicate better. It holds a number of competitions each year to honor researchers who excel in communicating science, including a prize for the best science-related article written by a young scientist, and a lecture series in which five young scientists who have been deemed "excep-

tional at communicating their research to a general audience"
are invited to make highly visible presentations.

Since 1987, the association has also placed up to ten scientists a year in short-term journalism fellowships with British
media outlets. In 2004, the association conducted a survey of
past fellows to determine whether this program was improving
relations between the clashing cultures.[8] Of the seventy respondents (about half of past fellows) only 30 percent had a "generally
positive" perception of science journalists before the program.
(The rest had "generally negative" or "neutral" attitudes.) After-
ward, 77 percent had a higher opinion (most of the past fellows
whose views were unchanged had positive views to start). A
focus group conducted as part of the evaluation showed that
at least some of the adverse views of science journalists were
based on the second-hand opinions of the scientists' colleagues,
not their actual experience. Three-quarters of the respondents
transferred the skills and experience they had acquired to colleagues by giving talks, drafting written advice, or by some
other means. Such findings give hope that programs that try to
bridge the divide have promise. They also bear out the research
of Sharon Dunwoody, who has found that scientists who have
more contact with journalists have more positive attitudes
about journalists and journalism.

The experience of Andrew Derrington, a brain scientist at
the University of Newcastle Upon Tyne and a British Association
for the Advancement of Science Media Fellow in 1994, illustrates
the effectiveness of this approach. Derrington was awarded a
summer fellowship to report on science at the *Financial Times of
London*, and the collaboration was so productive that he wrote a
science column there for the next five years and came away with
heightened respect for journalists. On Derrington's first assignment, the secretary of a scientist he hoped to interview informed
him that if he made a request in writing, the researcher could
make time within the subsequent three weeks. "I only wanted

five minutes," Derrington says, still smarting from the slight. Given the constraints of space on the page and time, he says he's concluded that journalists generally do a good job. To scientists who gripe about the quality of the reporters' craft he says, "If you had to explain your research in 10 words could you say it righter than the journalists do?"

THE TRUTH BEHIND THE STEREOTYPES

Unfortunately, programs like the British and American media fellowships only help a small number of scientists each year. And strained relations between scientists and journalists are often not just a product of simple misunderstandings. There is evidence to prove that some science reporters truly are ill informed and that some scientists really do have trouble speaking in jargon-free prose. The *Worlds Apart* study, for instance, showed that most journalists thought their own colleagues were generally ignorant of how the scientific method works. By the same token the report showed that more than 70 percent of scientists agreed that their profession often has trouble communicating in plain English.

Some communications researchers have tried to learn just how good (or bad) journalists are. In 1974, James Tankard, a journalism professor at the University of Texas at Austin, and Michael Ryan, a professor at Temple University, published a groundbreaking study on accuracy in science reporting.[9] Tankard and Ryan sent copies of newspaper articles on scientific research to the scientists who had conducted the research in question. Using a checklist for various types of errors, the scientists combed the articles for mistakes. The journalism researchers discovered that only 8.8 percent of the articles were error-free, a figure considerably lower than the 40 to 59 percent accuracy range previously published for general news stories. Later research reported lower error rates for science stories and raised some questions about what it means for scientists to judge the accuracy of popular articles summarizing their own work. Subsequent studies of accuracy in science reporting also showed that some of the "errors" identified by scientists were actually not incorrect facts but the perception that coverage was not a *complete* enough representation of research results (e.g., lacking an explanation of research methodology or important qualifications).

Sharon Dunwoody says this research on accuracy focuses too heavily on what she calls "technical accuracy," such as whether news stories mentioned coauthor, cited numbers correctly and the like. She prefers to focus on what she refers to as "communicative accuracy," a measure of whether an average reader of a news story (as opposed to a specialist) understands the main thrust of the scientific research being reported on. She says although there is not much research on this topic, there is some evidence that, by the standard of communicative accuracy, the majority of news stories are accurate. Nonetheless she concedes that "journalism is filled with errors large and small" (which she is "fanatical" about eliminating from the work of her journalism students).

One reason for errors and other shortfalls in science reporting could be inadequate training of journalists. Deborah Urycki, a medical and health writer, lays part of the blame at the feet of journalism departments for failing to give students the training they need to decipher accurately and translate scientific research. As a journalism graduate student at Kent State University, she conducted a study of the ability of seventy-nine journalism majors and minors to write articles summarizing scientific papers on "noteworthy" topics "of interest to the general public."[10] Urycki asked each of these journalists-in-training at three Ohio universities to write an article suitable for high school graduates, and a scientific panel graded the results. Not surprisingly, students who had taken more writing courses wrote better articles. The panel also found, however, that *all* of the students produced papers riddled with flaws, regardless of their interest in science, the number of years they had been in college, or any of a number of other factors considered.[11] The panelists' withering comments paint a distressing portrait of America's future reporters: "Article was hopelessly muddled"; "Abysmal writing, by turns incomprehensible and overly simplistic"; "Reproduced phrases from the article, but without total comprehension."

Sheila Tobias, who has written numerous books on science and math education, says that college education "weeds out" all but the most promising young scientists from undergraduate science classes, leaving the remaining students with little exposure to or interest in science. As a result, she says, large numbers of liberal arts undergraduates, including pre-journalism students, complete college convinced of three things about science: "First, that they're not very good at it. Second, that they don't like it very much. And, third (in defense of bruised egos), that science is irrelevant to their future."[12]

Journalists who specialize in science have greater interest and training in science than most reporters. They may also do a better job simply by virtue of the experience they accumulate with time. However, the number of science reporters on newspaper staffs is decreasing, and most science news is written by generalists, who generally don't have time to develop much of an expertise in any single branch of science. In broadcast news, where there are hardly any science specialists, the situation is worse. Clearly, improvements in the relationship between science and journalism require more than just better understanding. Journalists who report on science need better training. (Suggestions for how to obtain this training are included in the Resources at the back of this book.) Scientists, for their part, must improve their ability to explain their research to people with little technical background. Chapter 4 will help you talk to reporters in a way that conveys your message without talking down to them or oversimplifying. If you already have these skills, you may still want to use them more often.

THE BENEFITS OF A HEALTHY RELATIONSHIP

There are many reasons why, as a scientist, you might want to talk to the media. Like everyone, you probably have a variety of motivations. Perhaps you want to do something about the quantity and quality of science news. Perhaps you want to advance your career, or feel duty-bound to make science more accessible. Perhaps you aren't satisfied with having the university press office issue a press release that might be overlooked. Perhaps you want to be more involved to make sure your research is accurately portrayed. Perhaps you want to share your general scientific training and special expertise, rather than knowledge gained from a particular research project, to help public understanding of important policy issues. Or perhaps you are like ornithologist Bill Evans, who spent seventeen years prowling migratory-bird rest stops recording an audio catalog of elusive night calls. When the job was done, Evans wanted to share his results with the world. He created his own website and welcomed a journalist (Grossman) on one of his field research trips. For some, the only motivation needed is a desire to share the love of your work and the voyage of discovery with others.

Regardless of your motive, anecdotes and published research both show many tangible benefits (though not always predictable) from publicizing your research.

Public Health A 1991 study in the *New England Journal of Medicine* showed the impact of the media on medical researchers' familiarity with important scientific advances.[13] The authors found that journal articles covered in the *New York Times* were more likely to be cited subsequently in the medical literature. Other research has found that media coverage of the results of drug trials can influence the public to alter profoundly its choices of therapies. A 2004 article in the *Annals of Internal Medicine* found a link between widespread media coverage of a study questioning the benefits of hormone replacement therapy and a subsequent 30 percent decline in the number of women receiving this treatment in the following several months.[14] (The authors said the change was too abrupt to have resulted from the advice of doctors who read about the research in the *Annals*.)

In early 2005, Professor Richard Wurtman, director of MIT's Clinical Research Center, called the university's News Office about a paper he was preparing to publish in *Sleep Medicine Reviews*. Wurtman studies how food and drugs affect the brain, and his new paper, a "meta-analysis" of seventeen previously published studies, confirmed that the hormone melatonin can help people with insomnia.[15] The study showed that very small doses of melatonin help people fall asleep faster and return to sleep more quickly if they wake during the night, a problem many older people have. Elizabeth Thomson, assistant director of the MIT News Office, says the study is important because earlier research had raised questions about melatonin's efficacy. But these earlier studies involved much larger doses of melatonin, causing receptors in the brain to become unresponsive to the hormone and exposing people to possible side effects associated with high doses. "Professor Wurtman felt strongly," says Thomson, that "melatonin had gotten a bad rap. A lot of people could benefit from it."

After visiting Wurtman in his lab, Thomson and her staff crafted a publicity campaign for the research. Initially they tried offering it as an "exclusive" to two newspapers, neither of which took the bait. So they switched to Plan B, and sent a press release to 450 journalists around the world. In an interview shortly after issuing the announcement, Thomson reported that both the *Washington Post* and *Scientific American* were doing a story on the findings. "I'm ecstatic," she says.

Why go to all this trouble? Thomson says millions of sleep-deprived people could benefit from using melatonin, but neither they nor their doctors might ever hear about Wurtman's research if it were published in an obscure journal and nowhere else. "How many people read *Sleep Medicine Reviews*?" she asks rhetorically. Thomson adds that the university also benefits from touting the accomplishments of Wurtman and other faculty members. Every time MIT appears in the media, she explains, the job of recruiting top-notch faculty and accomplished, "well-rounded" students becomes easier. "The more you hear stories in the news with cool research, the more it benefits the institution." Thomson admits that Wurtman and MIT both also stand to benefit financially from publicity about Wurtman's findings since he and the university have jointly patented a low-dosage preparation of melatonin. Another financial benefit of publicity: "It makes funders happy," Thomson says. "It helps get more money."

Funding Publicizing its scientists' work is a job MIT takes very seriously. In her newly renovated News Office, Elizabeth Thomson oversees a staff of more than a dozen people devoted to burnishing the image of the university and its researchers. When an MIT professor wins the Nobel Prize, Thomson delegates someone to field and prioritize press calls. When a newspaper or television show wants to do a feature on the university, she sets up interviews. And when a scientist makes a discovery or publishes an interesting new paper, she sends a press release to some of the thousands of journalists in her database.

In fact, MIT's News Office produces as many as a dozen press releases every day, touting everything from breakthroughs in astrophysics ("Spacetime wave orbits black hole") to practical discoveries ("Plastic helps monitor pollutants") to campus buffoonery ("Dueling deans test their culinary skills at student event"). For the most telegenic stories the office also spends thousands of dollars producing videotapes for television producers.

The scale of this operation may not be typical of most American colleges; MIT is, after all, one of the world's leading science and engineering universities. (Grossman also happens to be a graduate of MIT, though he strives to be unbiased about his coverage of the university and its staff members.) Nevertheless, in terms of its desire to brag about itself, MIT is no different from hundreds of other institutions of higher learning in the United States. About two-thirds of all research and development dollars invested in such institutions come from federal and state coffers.[16] No wonder Harvey Leifert, the public information manager at the American Geophysical Union in Washington, D.C., says that scientists "have an obligation not just to the scientific community but to a wider audience to tell what they are doing and its significance."

The federal government spends around $118 billion a year on research and development, of which some $19 billion is destined for academic institutions.[17] State and local governments, industry, and other sources pitch in approximately $13 billion more.[18] So it's not just good form for academic scientists to let the public know what all that money buys. It's also good politics. Priorities change from year to year as governors, presidents, the Congress, their constituents, and others rethink how to allocate money. Moreover, legislators and administrators at the federal, state, and local levels consider new laws and rules that could affect scientific research every day. Everything from tax policies to hazardous waste laws to rules about stem cell research could have serious consequences for the scientific community. Scientists would be well advised to remind such people often of why scientific research matters.

Funding agencies also generally want researchers to publicize their findings. The National Science Foundation, for example, not only requires grant recipients to do good science but also urges them to share their results with the public. The agency tells its proposal reviewers to consider the "broader impacts" of each grant request. Among other criteria, the reviewers are instructed to evaluate such impacts as whether the results will "be disseminated broadly to enhance scientific and technological understanding?"[19] When it comes time to renew a grant, therefore, the agency is saying essentially that they will look favorably on researchers who can obtain press coverage for their research.

Career Advancement Lee Frelich, a researcher in the Department of Forest Resources at the University of Minnesota, concedes that although he doesn't choose research topics based on the potential for press coverage, there are "selfish reasons" for getting media attention when the results are in. He says his numerous appearances on television and in newspapers and magazines have given him greater stature in the eyes of the school's administration. This high profile has made it easier to get help from higher ups in setting up meetings with major donors and securing new office space. He admits that he does not necessarily do better science than a colleague down the hall who hasn't appeared in the news, but "that's the way the world works." Microbiologist Jed Fuhrman adds, though it is rarely acknowledged, that a scientist's stock at funding agencies goes up when an article about his lab gets good publicity. "My grant officer at NSF thinks much better of people who get things written up in the *New York Times*," he says.

Whatever your motive, the key to getting good coverage of your ideas and work is knowing how to attract the media's attention and then, once you have it, communicating effectively. We'll discuss how to accomplish these tasks in chapters 4, 5, 6, and 7. But first, in chapter 2, we'll warn you of the pitfalls you could encounter in your relationship with the media. In chapter 3, we'll explain some of the essential similarities of, and differences between, scientists and journalists.

CHAPTER 2

Hope for the Best, Prepare for the Worst

E VER HAD A nightmare where you're speeding down the highway and suddenly the pavement ends and you're hurtling over a cliff and way below you see foamy waves breaking on boulders? For some researchers, that's what it's like to get a call that begins, "I'd like to interview you." "It is risky," says Neal Lane, an atomic and molecular physicist and former administrator of the NSF who encourages scientists to be more proactive with the media.

Scientists can get burned when talking to the media by being misidentified, misquoted, quoted out of context, sensationalized, or even ridiculed—problems we hope to help you avoid. Tom Lovejoy, president of The H. John Heinz III Center for Science, Economics and the Environment and former chief biodiversity advisor to the World Bank, says you must still publicize your results outside of the scientific community despite the risks. "We are members of society and privileged to know and understand this complex and highly sophisticated subject matter that is an important underpinning of society," he explains. "Every scientist has a responsibility to communicate with the public."

In accepting this responsibility, we suggest you approach the media with an old exhortation in mind: "Hope for the best, prepare for the worst." In this chapter we'll look at what the worst-case scenario might be, and in the chapters ahead, help you establish effective, trouble-free communications with reporters.

THE SKY IS FALLING!

Scientists speak precisely and are controlled, unemotional, and humorless. That's what the stereotypes say, anyway. Scientists are also typecast wearing lab coats, dorky horn-rimmed glasses, and (at least in the movies) pocket protectors. These widely held images are, of course, at best exaggerations. But one thing that is certain about all scientists is that they will not accept anything as the truth unless they have the facts to prove it. This may explain one of the more common reproaches cast upon journalism by scientists. In most contexts "sensational" is a compliment. But when a scientist uses the word to describe a story about his work, it's probably with disgust, not delight. It may be, as is often the case, only the headline that qualifies as sensationalism. Or the entire article might be riddled with purple prose and exaggerations. Either way, your professional insistence on accuracy will probably be offended. Nonetheless, you may, like Lee Hannah, want to view some media distortions with equanimity.

Lee Hannah is a scientist at the biodiversity advocacy group Conservation International. He has had an unusually heated run-in with exaggerated headlines and stories hyping research results. He and more than a dozen co-authors were taken to task publicly for inaccuracies in a global spasm of reporting following the publication of a paper of theirs in *Nature*. The article, titled "Extinction Risk from Climate Change," appeared in early January 2004.[1] It contained an estimate that between 18 and 35 percent of the world's species could be "committed to extinction" by 2050 due to habitat loss caused by global warming. In the subsequent days and weeks, scores of stories on the research were published and broadcast by many of the world's major newspapers and broadcasters, including National Public Radio, the *New York Times*, the *Washington Post*, the *Los Angeles Times*, and the *San Jose Mercury News*.[2] The headlines of many of these reports seriously misstated the conclusions of the paper; a typical example was the *San Francisco Chronicle*'s "Dire warming warning for Earth's species: 25 percent could vanish *by* 2050 as planet heats up, study says."[3] The original scientific paper did not actually say that even a single species would be extinct by 2050. The paper and press releases distributed by the research team stated that the research estimated the number of species that could be *committed* to extinction by accumulated climate warming through 2050, not how many species would die out by that date. "Information is not currently available on time lags between climate change and species-level extinctions," cautioned the paper, "but decades might elapse between area reduction (from habitat loss) and extinction." Nonetheless, numerous newspaper headlines incorrectly reported that the researchers expected massive extinctions *by* 2050.[4]

In a vituperative letter subsequently published in *Nature*, Richard J. Ladle and several colleagues at the School of Geography & the Environment at Oxford University described "damaging simplifications" appearing in "sensational" news reports about the extinctions paper. The letter's authors had found errors in twenty-nine newspaper articles published in Great Britain, leading them to conclude that press coverage was "highly inaccurate."[5] The British team blamed the misrepresentations on the original authors.

Lee Hannah, one of the original paper's co-authors, and Brad Phillips, then a public affairs official at Conservation International, responded forcefully to Ladle's rebuke in another letter to the editor in *Nature*. They asked, rhetorically, whether *Nature* should have eschewed publicity for the research, as the Oxford team's letter appeared to suggest.[6] "We don't believe so," they wrote. Though they said they, too, deplored the inaccuracies,

Hannah and Phillips said that such mistakes were an unavoidable risk of publicity. On balance, they wrote, "[b]reaking through a U.S. media climate often dominated by news of war, terror or the latest celebrity escapades is a victory."

It could be that the damage done by the media's errors was outweighed by benefits of getting coverage of species extinction; that is for each individual to decide. On the other hand, it could be that the misrepresentations could have been avoided (or at least minimized) in the first place. After all, the difference between "by 2050" and "years or decades after 2050" might have seemed to at least some of the reporters like an unimportant fine point. Likewise, the use of the word "committed" might not have been considered to carry a critical distinction. And the authors' attempt to clarify the limits of their research (contained in the sentence "Information is not currently available on time lags between climate change and species-level extinctions, but decades might elapse between area reduction (from habitat loss) and extinction")—was not as direct as it could have been.

Clearly, the scientists forgot who their final audience was—not other scientists or even educated reporters, but journalists with little scientific background, and the public. As we discuss later in the book, once you know how to talk to the public *through* the media, press coverage of your research will be more accurate and cases like this will have much greater odds of being committed to extinction.

JOURNALISM FOR PROFIT

Can Prozac cause cancer? You would probably think so if you read "Scientists find Prozac 'link' to brain tumours," a headline that appeared on the front page of the March 26, 2002 edition of the *Independent*, one of Great Britain's major daily newspapers.[7] The paper takes its science-related coverage seriously, featuring weekly health and science sections and several environment stories every day. Yet it ran a story with inaccuracies that needlessly raised fears and concerns among its readers. The scientist responsible for the research could have prevented the blunder.

The Prozac story must have been alarming to many people, including what was most likely thousands of readers taking Prozac and related antidepressants. The story began, "Scientists have discovered that Prozac, the antidepressant taken by millions of people around the world, may stimulate the growth of brain tumours by blocking the body's natural ability to kill cancer cells."[8]

Dr. John Gordon, the principal investigator of the tumor research wanted publicity for his research, but he never anticipated that the reports would worry anyone. The Birmingham University immunologist was employing Prozac and other related antidepressants as tools to study the effect of the natural hormone serotonin on cancer cells. He had previously shown that when mixed in a test tube with serotonin, cancerous blood cells from the lymphoma family of cancers become deranged and commit a sort of corpuscular suicide called apoptosis. Gordon wanted to understand the mechanism behind this effect in the hope of discovering novel cancer treatments. As for the role Prozac played in his research, Gordon took advantage of the antidepressant's ability to block serotonin from entering brain cells. He and his team of scientists had thought that if they could use Prozac to stop serotonin from entering the lymphoma cells in their test tubes, they could determine whether the hormone causes cells to self-destruct by some action mechanism in the cell's interior rather than something on the cell's outer surface. When the Prozac did indeed halt serotonin's effect on the lymphoma cells, the researchers had their evidence that the hormone's anti-cancer effect does take place inside cells, not on the exterior surface. It was a small but important step in the search for better cancer treatments.

Shortly after his results were published in the journal *Blood*, Gordon received a call from Steve Connor, the science editor of, and a leading reporter for, the *Independent*.[9] An enterprising and experienced journalist who had written thousands of science articles during a career spanning twenty years, Connor had been alerted to the *Blood* paper by a pharmacologist who had suggested the Prozac-might-cause-cancer angle. As Gordon recalls the interview, after several routine questions, Connor tossed him an unexpected zinger: if Prozac blocks serotonin from destroying lymphoma cells in a test tube, could it cause cancer in the normal brain cells of people using it to treat depression? Gordon says he thought for a moment before responding, "Anything is possible." He added, however, that to make the conceptual leap from lymphoma cells in a test tube to brain cells in the body was "to go out on a limb"—beyond what the evidence justified. For instance, the effect of Prozac on lymphoma cells could be different than on brain cells. Or the drug might behave very differently in a highly complex organ like the brain than in the controlled environment of laboratory glassware.

When Gordon saw Connor's article the next day, he was concerned that some of the millions of people taking Prozac and related drugs for depression might abruptly discontinue treatment, risking a disastrous resumption of depression or some other mental health issue the medicine was treating. It wasn't long, as the scientist puts it, before "the shit hit the fan." Apparently touched off by the *Independent* article, the Prozac-might-cause-cancer hypothesis spread like a turbocharged virus to papers around the world. Gordon went without sleep for the next forty-eight hours doing "damage limitation" with non-stop interviews across Europe and in the United States, Australia, and South Africa. It took two weeks before the media frenzy subsided.

The immunologist believes the affair was a clear case of a circulation-boosting, goosed-up story. "It made a good headline for them, but it had an impact that could have caused serious problems for those taking and needing the drug," Gordon says. Journalist Connor is unrepentant. Referring to Gordon and the Birmingham University press office, he says the imbroglio was "a mess of their own making." The reporter says he called the scientist in the morning the day before the story was published but Gordon instructed him to make his inquiry through the press office, which he did. Connor wrote his story as best he could with just a copy of the scientific paper while waiting for the press office to arrange an interview. He says the immunologist didn't call back until the evening, by which time the deadline for the following day's paper had nearly expired. The journalist says he was able to shoehorn some last-minute quotes into the story, but it was too late for him to retool the article's basic thrust. To the journalist, the case is "a prime example of what happens when the press office decides to gag one of its own scientists." Connor does, however, wish the article had a different headline and first paragraph. He says he did the best he could with the information and time he had. "Journalism," says the reporter, "is not an exact science."

Sue Primmer, Birmingham University's director of communications, admits that Connor stumbled onto John Gordon's research before her team was ready to respond. "His timing didn't drive the story," she says. "The driver was John [Gordon]'s research." She regrets that Connor "took a lot of flak" for his story since it was "80 percent right," and most of what was wrong was the title, which he didn't even write. That sensational headline propelled what might have been a very small media event into a worldwide sensation and, ironically, Primmer says (in terms of getting publicity for Birmingham University), "did us an enormous favor."

What is a scientist to make of all this? No researcher wants publicity like this. First, the brouhaha around Gordon's research was easily avoidable. Given that the cancer research had already been published in the journal *Blood*, Gordon and his press office should have had a plan in place to respond to reporter calls promptly. A whole workday should never have passed before the reporter interviewed Gordon. Second, never ad lib, especially with a question as provocative as the one Connor tossed Gordon. ("Anything is possible" is a risky answer no matter what the question.) The chapters ahead provide more interview techniques and strategies to ensure that you know how to respond effectively to reporters' questions.

What happened to three University of Edinburgh scientists in 1997 illustrates a different sort of problem and requires a different solution. Ian Deary, Martha Whiteman, and F.G.R. Fowkes published a paper in the *Lancet* on the relationship between personality and heart disease. The study was based on interviews with 1,600 men and women over five years.[10] Among the researchers' findings was that women who are more "submissive" are up to 31 percent less likely to suffer heart attacks. The authors fretted from the start that the press might misunderstand or misconstrue their research, which had several elements that newspapers—especially sensational tabloids—find irresistible. In particular, the study dealt with the risk factors associated with a dreaded disease that causes sudden death (heart disease), and the research could be interpreted as lending credence to outmoded but still widely held gender stereotypes. Accordingly, the researchers drafted their press release with care, or so they thought. Most importantly, the Edinburgh scientists wanted to be sure that the operative word, submissive, was understood correctly. For the purposes of their study, people with higher submissiveness were those who voluntarily permitted others to take the lead in interpersonal decisions. Thinking they were playing it safe, the researchers substituted what they considered a relatively neutral term, "meek," for what they considered the more charged "submissive."

Had the scientist team realized that even the word "meek" was a highly loaded term for the public, the media debacle that followed would likely have been avoided. Instead, the scientists were blindsided. When the embargo on their research was lifted, the headlines and, in some cases, the body of articles themselves were loaded with gender stereotypes in no way substantiated by the research. For instance, the story in the *Daily Telegraph* ran under the headline "Put down that rolling pin darling, it's bad for your heart."[11] Though the scientific paper itself said nothing about gender-based roles, many of the newspaper articles implied, if obliquely, that the research was a scientific endorsement for the old-fashioned division of labor in the family. In a postmortem subsequently published in the *Lancet*, the researchers spoke of the pain of "watching data from thousands of patients collected over several years trivialized, distorted, and used in some outlets to support a set of misogynistic attitudes."[12]

This is a shame. If the researchers had worked harder to find a more suitable alternative to submissive, the stories would probably have been more accurate. That's why, as we discuss later in the book, scientists should test their language on non-scientist friends at a party or on disinterested relatives over the phone.

A 2002 editorial in *Nature* on media peccadillos advised scientists not to succumb to the temptation of becoming embittered by faulty reporting or hyped headlines.[13] Instead, the editorial suggested scientists should take a page from the playbook of politicians, who are frequently misrepresented yet who know that retaliating against the press or withdrawing from publicity is counterproductive. "[T]hey know that attacking journalists is a shortsighted strategy," the journal stated. "Instead they have become experts in rebutting inaccurate stories and imparting their own message." That makes a lot of sense, and we offer tips in this book on how to respond to bad stories. But follow our advice in the chapters ahead and you can avoid most of these issues in the first place.

THE QUOTABLE SCIENTIST

Sensational headlines and text are only one type of mishap that researchers can suffer when they step into what is sometimes the purgatory of the public spotlight. Another is being mis-quoted. The consequences of this misfortune range from the humorous to the deadly serious. Even correct quotes can cause heartache in some instances. Some scientists are angered when they discover that neither their words nor any of their thoughts make it into a story, sometimes after spending hours tutoring a reporter. Others discover, to their dismay, that their findings or their words have been twisted to support wacky theories.

Susanne Moser, a research scientist at the National Center for Atmospheric Research in Boulder, Colorado, has been inter-viewed about global warming scores of times by dozens of jour-nalists in every medium. Most press encounters go well, but every once in a while she has a humdinger. Take, for instance, an interview she granted a reporter in 2001. The journalist was writing about how President-elect George W. Bush might respond to global warming. The reporter had recently been assigned the environmental beat for the first time. "She didn't have a clue," says Moser, who spent two and a half hours bring-ing the journalist up to speed on the difference between the short-term fluctuations of weather and the long-term trends of climate, and on the theory of how the greenhouse effect causes global warming. Moser was livid when she saw the resulting story. The problem was the article had important factual errors. For instance, it conflated weather and climate. "It was as if we hadn't had the conversation," says Moser. What really galled her was a short paragraph in the middle of the article undercutting everything else. "A small number of scientists say the shift [in Earth's temperature] is cyclical," said the article, "and has noth-ing to do with human-caused interference." Moser points out that nearly every qualified scientist not on the payroll of a com-pany tied to the consumption or extraction of fossil fuels agrees that humans are a (if not the) major cause of global warming. She says undermining the consensus of scientists who believe human activities are warming the planet with the opinion of one or two skeptics does a disservice to the public, which is forced to make sense of the jousting "experts." "It's a huge problem," she says.

These details are important. On the other hand, are scientists sometimes guilty of over-reacting? Four years after her unfavorable experience with the reporter writing about the president elect's expected policies, Moser says the global warming story doesn't bother her as much as when it first came out. In retrospect, she now calls the incident a "perfect example of a scientist—very interested in the details and in getting it right—having a problem with imprecise, unsophisticated, and misleadingly 'balanced' coverage." Andrew Derrington, the brain scientist mentioned in chapter 1 who briefly exchanged his academic post at Great Britain's University of Newcastle Upon Tyne for a science journalism position at the *Financial Times of London*, says scientists often hold media reports to a standard that articles for non-experts cannot—and should not—meet. Of course, everybody agrees that when a journalist cites a specific number, the correct one should be used and that names should be spelled correctly. On the other hand, including every detail of a scientific study's research design, the history of previous studies, and a thorough critique from other researchers (not to mention a complete list of coauthors) is not even a desirable goal in a popular medium like a newspaper. Journalists are limited by the amount of space (or, in the case of broadcast news, time) available to them, and also need to avoid boring or confusing their audience. "If there were a paper written the way they would like it," Derrington says of some scientists critical of the press, "nobody would read it." We in no way wish to excuse sloppy practices. On the other hand, researchers should keep a sense of proportion, like Lee Hannah. Though displeased by incorrect articles about species extinction, Hannah considered the errors an unfortunate part of the price to pay for getting extensive attention for his research.

Although reporters do sometimes get quotes wrong or quote out of context, interviewees should realize that in journalism, a quotation is considered an inviolate truth. An article can be slanted one way or another depending on many factors, but the words inside quotation marks are considered sacrosanct. Of course, even this reverence for precision permits some exceptions. For instance, some reporters will clean up an interviewee's grammar as long as it is trivial. Despite their best intentions and most careful efforts, though, journalists do sometimes make mistakes and quote incorrectly. Scientists also make mistakes and later regret their words. Unless you tell the reporter that you are not speaking on the record (and even then there are risks), everything you say in the presence of a journalist could be quoted— so choose your words carefully. Wildlife biologist Warren Aney learned this lesson the hard way. For more than a decade until he retired in 1993, Aney was a regional supervisor for the Oregon Department of Fish and Wildlife. He was responsible for a region in northwestern Oregon that included one-fifth of the state, an area almost the size of West Virginia. Aney had studied journalism before he became a wildlife scientist, so he considered himself media-savvy. If there was a problem in his district like an issue with a fish hatchery or insufficient winter precipitation, he'd call local papers before they called him. If he got an unsolicited press call, he didn't react defensively; he considered the call "an opportunity to educate the public." As a result, Aney was accessible to his subordinates and the press alike, making it clear that reporters with questions should talk to him, not some distant public affairs department. In other words, Aney developed into a good source for reporters, a step we recommend in chapter 6.

Yet even a media-savvy scientist like Aney can sometimes slip up. One memorable instance occurred while being interviewed for an article about plans to reintroduce bighorn sheep into the Eagle Cap Wilderness in the Wallowa Mountains. At the time, Aney was upset with the U.S. Forest Service for "dragging its feet" on the removal of domestic sheep from the area. The federal agency permitted ranchers to graze sheep in parts of the territory where Aney wanted to introduce the bighorn, but domestic sheep are carriers of a pneumonia-like disease that infects and kills the wild animals. After asking some general questions about the program, the reporter asked Aney what was the most important obstacle to reintroducing the bighorn. It was a perfectly fine question, but Aney had gotten too comfortable with the reporter. Letting down his guard for a brief instant, he said exactly what was on his mind: the U.S. Forest Service.

When the bighorn article appeared in print, Aney was quoted about Forest Service obstructionism. "I got too accurately quoted," says Aney with a sense of humor that only comes with the passage of time. He was subsequently called into the office of the top Forest Service official in the region, who "let me know he was dissatisfied" with Aney's frank assessment. Aney says in retrospect the federal agency may have drawn out the bighorn introduction in retaliation for his blunder. "What could have happened in two years," he says, "took 10 years instead." The lesson? Don't say it if you don't want to read it over breakfast.

David Pisetsky, a rheumatologist at Duke University Medical Center, says medical researchers sometimes make exaggerated or unwarranted claims that, when repeated by the press, are more than just embarrassing. They also cause anguish for patients and doctors alike. He recalls once spending a day tracking down a news report of a "breakthrough" that a terminally ill patient thought might cure him. It was an unpleasant experience, because he had to explain that the research the desperate patient had heard about in the news was still very preliminary. Another time Pisetsky had to explain to his own mother, who has osteoarthritis, that a "cure" she had heard about on the news "isn't there yet." Pisetsky encourages medical practitioners to consider carefully how they describe their work to reporters lest they "raise hope and then there's nothing there."

ERRORS OF OMISSION

Scientists with grievances against the media typically refer to a variety of errors of commission, many of which have been described earlier in this chapter (such as sensational headlines and incorrect quotes). The media also stand accused of less visible but equally serious errors of omission. Many communications studies show that scientists often consider news stories deficient because they don't contain enough (or any) information about research methodology and previous research and omit qualifying statements, such as the preliminary nature of conclusions, or the views of critics.[14] A more subtle error of omission is stories or topics that don't get into print or on the air. The unwritten norms for what is newsworthy, daily deadline pressures, commercial demands on publishers, and a number of other factors explain why the media favor certain kinds of stories at the expense of others. For instance, all other things being equal, stories get published or broadcast if they concern some notable superlative (the newest, smallest, coldest, etc.), if they are counterintuitive or unexpected ("man bites dog" stories), or if they have a strong emotional dimension, such as a previously unknown and dire threat to health or wealth or, conversely, a dramatic cure.

Vikki Entwistle reported in 1995 in the *British Medical Journal* that medical journalists at four major British newspapers scanned the tables of contents of the *Lancet* and the *British Medical Journal* each week for "key" diseases, such as cancer and AIDS, and for eminent authors.[15] The journalists told Entwistle, then a student at London's City University, that they were more likely to cover stories about "common and fatal diseases; rare but interesting or quirky diseases; those with a sexual connection; new or improved treatments; and controversial subject matter or results." These findings were largely confirmed and expanded upon by a study of two years' worth of papers published in the *Lancet* and the *British Medical Journal*. The study found that newspapers are more likely to publish research with "bad news," such as a study finding that infants sleeping on a sofa with their parents are at increased risk of sudden death, than "good news," such as a paper showing the positive benefits of jogging on mortality. Research on women's health, cancer, and reproduction received disproportionately more coverage in newspapers, whereas research on diabetes, heart disease, and the elderly received disproportionately less coverage.

Capricious (though explicable) practices like these explain in part why John Gordon's paper in *Blood* provoked worldwide headlines about the possible risks (bad news) of Prozac and related antidepressant use, while a team of researchers at the University of Toronto who subsequently published a paper in the *American Journal of Epidemiology* finding no link (good news) between such drugs and the incidence of non-Hodgkin's lymphoma received little, if any, news coverage at all.[16]

It's worth noting that while the media usually report on politics as a continually unfolding story, they generally cover science as a series of discrete, disconnected events or "breakthroughs" or, as Tineke Boddé writes, as "snapshots" rather than a "motion picture."[17] The unfortunate practice of treating science this way encourages the incorrect belief that scientific results are final, immutable end points. Thus, when new research modifies prior findings, the public feels misled, as if scientists keep pulling the rug out from under them. Another unfortunate consequence of portraying scientific research as a sequence of bolts from the blue is that there is no apparent need to update a story as new research advances a field incrementally, leaving the public seriously misinformed.

It's just one more reason why, despite the risks, scientists should do more to influence media coverage of scientific issues.

CHAPTER 3

Why Reporters Do What They Do

J OURNALISTS AND SCIENTISTS frequently enjoy fruitful working relationships, but just as often they find themselves at odds. As we've already discussed, journalists are often poorly versed in some or all of the disciplines on which they report, and scientists often fail to explain their work effectively. Though these problems are serious, they can be alleviated in a number of ways as discussed in chapters 4, 5, 6, and 7.

The economics of print and broadcast journalism and the increasing use of the Internet by the public as a news source put new stresses on science journalism that add to its shortfalls. Many of these problems are not likely to be seriously addressed, much less solved, anytime soon. As a scientist, you may not be able to change the culture of mass media, but you can help improve the way science is covered. If you understand what causes friction between you and your pen-, microphone-, and camera-toting counterparts in the media, you'll be better prepared to anticipate problems, helping to ensure better media coverage of science. At the very least, if you do experience trouble, you'll understand what went wrong.

SCIENCE NEWS VS. NEWS YOU CAN USE

The most important obstacle to scientists who want to publicize their ideas or research is a dearth of, and, in some cases, a decline in, the amount of science news carried by American news media. A study of the content of the nightly newscasts on the three commercial television networks shows that whereas the percentage of time devoted to science topics rose from 4 percent in 1977 to 11 percent in 2001, it declined to only 2 percent by 2003.[1] (Likewise, about 2 percent of front-page stories in a broad survey of U.S. newspapers concern science.)[2] Newspapers also increased the visibility of science in the last several decades, only to backtrack. In 1978 only one newspaper in the United States had a weekly science section (the *New York Times*), whereas in 1986 there were sixty-six U.S. papers with weekly sections.[3] Most of these sections have since either been retired or have been transformed into health, technology, or other more consumer-oriented sections.

The relatively small amount, and decline, of science in the news is due in part to an under appreciation of the value of science as a news topic. It is also affected by an overall downward trend in newspaper readership and network news viewership, and demographic changes in the audience for these media (specifically, the loss of adults between their mid-20s and mid-50s and college graduates, groups whose spending habits advertisers prize). The shrinking size and aging of the audience who use these media has led to fewer pages and fewer on-air minutes devoted to news of any kind. According to one study, the amount of time devoted to news (the "news hole") in a half-hour network nightly news segment declined from 21 minutes to 18.7 minutes in the twelve years between 1991 and 2002, an 11 percent reduction.[4] At the same time, opinion surveys suggest that the number of people specifically wanting to read science and technology news is declining. A 2003 survey found that 87 percent of Americans say they are "interested" in scientific discoveries, but this professed enthusiasm is belied by polls, in which people rank science below other news topics they follow closely.[5] Science dropped from sixth to eighth place in these rankings between 1996 and 2002.[6]

Partly as a result, the dedicated science reporting staffs of newspapers, already small, have shrunk. There is not one single network news producer or reporter devoted solely to covering environmental topics. Whereas networks used to have science and health reporters, now one reporter generally covers both of these topics. As mentioned above, most of the science sections launched by newspapers during the 1980s and 1990s are no longer being published, including those of the *Sacramento Bee*, the *San Jose Mercury News*, the *Dallas Morning News*, and the *Minneapolis Star Tribune*. Journalism professor Deborah Blum says television shows and daily papers today are generally reporting less pure science and more "news you can use," like mental and physical health or the weather.

On the bright side, the public is very interested in learning about practical applications of science. For example, the percentage of people who follow health news "very closely" trails only the percentage who follow crime and community news.[7] Health news ranks higher than sports, local government, national politics, and international news.[8] This interest has real repercussions. According to the 2003 survey of doctors conducted by Teresa Schraeder, news reports have a notable impact on the doctor-patient relationship. The overwhelming majority of the doctors responding to the survey reported that the media are a major source of medical information for their patients. And more than 80 percent of these health practitioners said that they themselves sometimes learn about medical developments in the popular press. More than a third alter clinical practices based on health reporting.

The reason that health and medical news is so popular is that it directly affects people. We will stress this point again and again, because even if your own work focuses on wildlife or the ocean or the stars, it is critical to talk about your work in ways that are relevant to people's lives—just like their health.

BAD NEWS

At its best, science journalism can inspire the public to appreciate the amazing complexity of nature and the joy of understanding it. It can give people power to improve their health and to protect their environment. It can also invite misinformation or even incite hysteria. Consider a case from 1990 that some journalism professors use to illustrate problems in journalism today.

"Welcome to Health Week. We have reports today on some new research into AIDS."[9] So began the June 2, 1990, edition of CNN's show *Health Week,* which went on to describe an experimental AIDS treatment called hyperthermia. This procedure, which is occasionally used to treat certain serious infections such as syphilis and some advanced cancers, involves draining a patient's blood from an artery, circulating it through a heater, and funneling it back into the body. The heat is intended to kill cancer cells and infections and to enhance the body's own defense mechanisms. The CNN report discussed the first (and, at the time, the only) use of such blood heating, several months earlier, for treating AIDS.

The patient, Carl Crawford, was described as having been severely ill and suffering from, among other things, Kaposi's sarcoma, a form of cancer associated with AIDS that causes black-and-blue splotches. Under general anesthesia, his blood had been heated gradually until his body temperature reached 108 degrees, a level far above what, under normal circumstances, would be fatal. Then, after two hours, Crawford was slowly cooled down.

While cautioning that a single case could not prove that a treatment works, and paraphrasing an unnamed doctor who questioned the therapy's efficacy, Dan Rutz, the CNN reporter, presented an upbeat picture of the procedure. Rutz stated that Crawford's Kaposi's sarcoma had begun to "heal and fade." Moreover, he said that subsequent "medical tests can reveal no signs of AIDS infection." Appearing at several points in the report, Crawford's two doctors, Kenneth Alonso and William D. Logan, Jr., then of the Atlanta Health and Lung Clinic at the small Atlanta Hospital, were cautiously optimistic. Mr. Crawford himself appeared in the report, exclaiming, "They can't say I'm cured, of course, you know, but I feel that I am cured, I really do."

Prior to the national CNN broadcast, the story had appeared in a report produced by WXIA-TV, Atlanta's NBC and CNN affiliate. The network's show sparked a nationwide flood of television, magazine, and wire service stories. Many reported the unproven treatment with more skepticism than CNN, introducing doubts raised by independent researchers near the top, rather than the bottom, of their coverage. Still, for nearly two months, during which time Alonso and Logan reported treating a second patient, the unproven heat treatment received mostly favorable coverage as a possible cure for AIDS. At one point, when the second patient was treated, CNN reported live from Atlanta Hospital, turning this early stage of medical research into a media event on par with natural disasters, plane crashes, and celebrity trials.[10] Other AIDS researchers were skeptical about the heating procedure and angry with the premature press coverage. But desperate AIDS patients were anxious to get in line to try what by most media accounts appeared to be a promising cure. The researchers reportedly fielded 1,000 inquiries a day.[11] A survey conducted by a New York doctor found that the overwhelming majority of doctors treating AIDS had been asked about hyperthermia treatment by their patients.[12]

Then, in mid-August, the third patient to receive hyperthermia treatment died. Not long after, a team from the National Institute of Allergy and Infectious Disease released the results of an investigation (begun before the death) of the research. The team concluded that "there appears to be no clinical, immunologic or virologic support" for using blood heating to treat AIDS.[13] The team found that Crawford never even had Kaposi's sarcoma in the first place—they speculated that he had been suffering from "cat scratch fever," a bacterial infection. The government investigators also concluded that improvements the Atlanta doctors reported in Crawford's condition were nothing more than a short-term response to antibiotics administered to prevent postoperative infections. The second patient never showed any signs of improvement. The investigators stated that there was no reason "for further human experimentation in this area at this time." Alonso tried unsuccessfully for years to obtain permission to conduct more human trials. His medical license was ultimately suspended by Georgia's State Board of Examiners for unrelated "unethical and dangerous" treatment of cancer patients.[14] Meanwhile, Atlanta Hospital closed its doors in late 1990, preempting state plans to revoke its operating license (for unrelated deaths caused by another doctor).

What happened? Why did CNN tout an untested cure by a little-known doctor at a faltering hospital, vainly raising the hopes of thousands of AIDS patients and their families? Though this may be an egregious example, it illustrates some of the shortcomings of American journalism, especially television reporting, today.

Follow the Money Although they may not know it or ever acknowledge it, journalists and scientists strive for fundamentally the same goals: to find and report the truth. These shared objectives ought to form a solid foundation on which to find common ground. What separates journalists from scientists and, to some degree, antagonizes them, are the divergent means the two professions use to seek, verify, and publicize the truth. One critical difference is in the ethos of their inquiries.

Science, in the words of *Silent Spring* author Rachel Carson, "is the what, the how, and the why of everything in our experience."[15] Scientists study problems that could confirm or disprove theories. They improve methodologies for making observations. They catalog phenomena and seek patterns in nature to make sense of undifferentiated facts. In contrast, according to Gary

Schwitzer, a journalism professor at the University of Minnesota, "journalism is driven by what is new." Schwitzer was the head of CNN's medical news unit in 1990 when the hyperthermia story was broadcast. He says CNN got the story wrong because its editors (not including Schwitzer, himself, as he was adamantly opposed to the story) were anxious to broadcast the first national show about the blood-heating procedure. "I was in favor of being slow and second and right," he says, but he was overruled. His superiors decided that a new AIDS treatment was news and they didn't want to be scooped. Schwitzer says that when he saw the show on the air, "It was like a saber had pierced me." He says there were many aspects of the research that should have caused the reporters to be suspicious. Atlanta Hospital was a fifty-bed community hospital under state investigation in the death of several patients. The doctors had no track record in research and had not yet been independently assessed (the National Institute and Infectious Diseases team had not yet started their investigation). "That's when I decided I couldn't work at that network anymore," he says.

News networks focus on what's new largely for financial reasons. Apart from nonprofit public television stations, which attract comparatively few viewers, television stations are companies expected to turn a profit. They do this by selling advertising. The goal, says Richard Wald, a journalism professor at Columbia University, "is not to inform you." Wald, a former president of NBC News and senior vice president of ABC News, notes that until the 1980s federal regulatory policies required television stations to temper their profit motive with a commitment to public service. But changes in Federal Communications Commission rules, deregulating the broadcast industry, and consolidation of station ownership changed that. Now, says Wald, "the basic outline of this whole thing is to make money." That has led to a slavish quest to elevate ratings, a measure of the percentage of household's viewing a show. For example,

WXIA-TV's first report on the hyperthermia research story was broadcast in late May during a "sweeps week," when television stations try to boost viewership with sensational stories. The number of viewers watching a station during a sweeps week (as determined by surveys sent to millions of Americans) is used to calculate advertising rates for the subsequent quarter. During these critical weeks, health stories are the honey for the news media's bees, viewers.

At least one commentator speculated that broadcasters overstated the significance of Dr. Alonso's heat treatment research merely to increase ratings.[16] This theory may not be completely accurate, but it does contain more than a grain of truth. Asked by a reporter from *Time* if broadcasting live from the hospital was an appropriate way to report on such unproven research, CNN's director of public relations at the time, Steve Haworth, said, "It depends on what is going on. We had no other breaking story that day."[17]

Often when television shows want to goose ratings, or when news magazines want to boost newsstand sales, they feature medical and health stories prominently. But reporters qualified to produce and write such stories are expensive, and most broadcast stations are trying to cut costs. Advertising revenue in television news is falling as viewers tune out and competition from the Internet increases. Daily newspapers and many magazines are straining under similar financial burdens. Media consolidation has compounded this problem by making more papers and television stations accountable to distant owners who have little, if any, emotional connection to the communities their media outlets serve. Today, newspapers accounting for 70 percent of the daily circulation in the United States are owned by just twenty-two companies; 23 percent of all radio stations are owned by just twenty companies; and just ten media companies reach 85 percent of the nation's households with 30 percent of all television stations.[18]

The profits these owners earn are actually sufficient to hire qualified reporters, but the companies expect huge profits, out of proportion with earnings elsewhere in the economy. This restricts the amount of money that can be devoted to the newsroom. For instance, despite declines in viewership, the profit margins of television stations are typically 50 percent and sometimes higher. Newspapers operated by corporate chains have 20 to 30 percent profit margins (in contrast to margins earned by independent papers of about 10 percent prior to the 1970s, when chain ownership began in earnest).[19] Even so, "owners of news organizations want higher profits," says Frank Allen, executive director of the Institute for Journalism & Natural Resources and a former environment editor and bureau chief for the *Wall Street Journal*. "Higher profits come with larger revenues and lower costs." As a consequence, he says, news organizations are investing in better ways to distribute their products while cutting news-gathering budgets. Richard Wald says the resulting loss of quality in commercial television programming contributes to a vicious cycle of declining viewership and lower advertising revenue that, over the long term, cannot be sustained.

These financial forces directly affect journalists. For example, since the average number of reporters in network newsrooms peaked in 1985, it has declined 30 percent, from 77 to 50 by 2002.[20] The average number of stories produced per reporter during the same period increased 30 percent, an increased workload that no doubt has had an impact on the quality of their work. In a 2003 study of the state of natural resource and environment journalism at 285 daily newspapers in the North American West, Allen reported that 83 percent of editors identified having too few reporters as the most important obstacle standing in the way of better coverage.[21] He also found that 80 percent of environment reporters at these papers are frequently diverted from their assignments to cover breaking news like fires, storms, or violent crimes. Other than the country's largest, most prestigious newspapers, most papers make do with at best a single environment reporter or nobody at all devoted exclusively to the subject.[22] The same pattern is repeated in print and television with health, space, science, and other technical topics.

Gary Schwitzer reported in 2003 that not one of the four network-affiliated television stations in Minneapolis had a full-time medical reporter (even though one of these stations reportedly had a profit margin of 70 percent).[23] Schwitzer systematically monitored the medical coverage of these stations for four months in the spring of that year, and in an article he published in the *British Medical Journal,* he reported that fifty-eight different journalists had filed medical stories for the four stations during this period.[24] "It's a terrible situation," says the journalism professor. "If you don't come in on any day owning an area, then you don't feel a sense of ownership." That, in turn, means reporters won't develop the specialized knowledge and contacts needed for top-quality journalism. Schwitzer says the problems caused by failing to assign specific reporters to these technical areas are compounded by small or nonexistent budgets for training and travel to conferences.

THE EASY WAY OUT

To compensate for the pressures they face in the newsroom, reporters have developed a number of shortcuts to make their jobs easier (often to the detriment of quality journalism).

Viewing Research in Isolation Journalism is rooted in a conceptual framework that is not always especially suitable for covering science. The earliest journalists in the United States reported on discrete events with relatively clear beginnings and endings such as court proceedings, crimes, fires, and the like. That is still their major role. Science, however, is a process, not an event (or, more colorfully, it is news that doesn't break, it oozes). In contrast with crimes and disasters it advances incrementally, without a clear beginning or end. It reflects our ever-changing state of knowledge about the world. With other topics, journalists generally wait for the story, or some identifiable aspect of the story (like the completion of a hearing in a criminal prosecution), to reach its conclusion before producing or writing an article. (Coverage of the gradual unfolding of a presidential electoral campaign is an exception that, though troublesome for other reasons, may be worth emulating for science reporters.) When reporting science, however, journalists don't have the luxury of waiting for the story's closing chapter, since the quest for knowledge goes on forever. "This can be frustrating for people who grew up on a diet of certainty," says Frank Allen, referring to journalists not used to covering science.

Gary Schwitzer compares science to a meandering river. Like a circuitous watercourse that moves one way then reverses course, research findings sometimes confirm and sometimes contradict previously accepted knowledge. Sometimes they point in completely unanticipated directions. Journalists, he says, tend to report on scientific research findings the way an explorer might report about a river if he or she could see only one small stretch of streambed. Reporters take the latest result and "offer that one thing as the gospel." As a result, a new pain reliever dubbed a "super aspirin" one week could be a "killer" the next. "That's why the public feels so jerked around," says Schwitzer.

LOOKING ONLY AT THE SURFACE

As chroniclers of politics, crime, disasters, and so forth, journalists have developed efficient techniques for learning and telling a story's "who," "what," "when," and "where." These techniques have been passed down in newsrooms from seasoned reporters to greenhorn copy boys, and in journalism schools as well. Frank Allen says that journalism is not nearly as good at describing the "so what?" and the "why"—attributes of a topic that take on critical importance in science-related stories. Take, again, the subject of global warming. It's a challenging topic for reporters and it is often handled poorly. The "who" is not your average villain (since it is everyone) and the "where" is not your average location (since it is everywhere). But what really matters with global warming is the "so what?"—the ecological, physical, and economic consequences of warming. To report this well takes familiarity with atmospheric physics, climate history, climate modeling, ecology, economics, and, if the story deals with solutions, politics. It also takes time. And time is scarce. Matt Hammill, a television anchor and environmental journalist in Illinois says when reporters at his station pitch a story, "The first question in the newsroom is can we do it in a day?" Not an attitude conducive to choosing complex topics. The second question, he says, illustrates the fierce competition for limited air time: "The reporter over there has a transvestite double murder, what do you have?" Hammill says in that environment daily news coverage can become little more than what he and his colleagues call "flash and trash," leaving an issue like global warming out in the cold.

In view of these shortcomings it's not surprising that reporting on all aspects of science is generally shallow. A survey of television news directors at 302 local stations conducted for the Radio and Television News Directors Foundation in 1995 shows that, as might be expected, television news segments on health and medicine focused on easy-to-report stories about "breakthroughs" and the like rather than more complex topics.[25] The news director at 96 percent of these stations reported running stories about health threats like diseases in the month prior to the survey. At 92 percent of the stations, the news director reported running stories about medical breakthroughs. In contrast, relatively few stations had run stories that dealt with how medicine was paid for, the quality of medical care, or how health care was administered. Only 30 percent of the stations had featured reports on how HMOs (then still a novelty) work, only 34 percent had reports on the relationships between doctors and patients, and only 58 percent had stories about health care costs.[26] A related survey of television viewers found that people were generally dissatisfied with what they considered superficial local reporting on health care. The report summarizing these results concluded that "local TV news must commit the time and resources to cover these stories in greater depth."[27] By all accounts the recommendations were not heeded.

FOLLOWING THE PACK

Gently poking fun at journalists' image of themselves as hardheaded skeptics, a reporter's joke goes: "If your mother says 'I love you,' check it out." Journalists are proud to tout their disinterested independence, yet much of what appears in print or is broadcast is a slight modification of what other journalists have produced elsewhere. The practice is so ingrained that journalists have invented their own disparaging term for it: pack journalism. Gary Schwitzer says that when he headed CNN's medical unit, reporters and editors were rewarded for rehashing what others had done, not for being original. "If they didn't have what the others had, you'd have to answer for it." He says this mentality is also one of the factors that helps explain why CNN followed the lead of its affiliate WXIA-TV on the hyperthermia story and why dozens of other print and broadcast outlets joined the pack after CNN's nationwide broadcast.

USING INTERVIEWS TO SUPPORT A PREDETERMINED ANGLE

Scientists complain that journalists sometimes seem to have written their stories before they even conduct their interviews, and that the purpose of the interview is merely to attach a name or a face to a preconceived opinion. We wish we could say that appearances can be deceiving, but this is actually common practice. As a working journalist (Daniel Grossman), one of us has done this many times. But as we hope you'll agree after reading on, the reality is not as damning as it sounds.

There are two fundamental dimensions of journalism: content and style. Content is the facts—what the story is about and how it is conceptualized. Style is how the story is told—the *Weltanschauung* of the topic and the persuasiveness of the argument. As veteran radio reporter Bruce Gellerman, former host of the public radio show *Here and Now*, says, content is the "meat" of the story, style is the "motion." The cable channel C-Span, which broadcasts unfiltered Congressional hearings and the like, is close to pure meat. So-called reality TV, in contrast, is virtually pure motion with little or no meat.

Good stories always have both meat and motion: meat is the message and motion is what makes the story engaging. Quotes or sound bites are put in stories mostly to add motion. Journalists generally do not use other people to say facts they could have said themselves. The voices of others make a story lively (as when an interviewee uses colorful language or speaks in an interesting accent or expresses strong emotions), add gravity to the story (as when the speaker is known to the listener or reader as respected or reliable), and lend credibility to particular points within a story (as when a scientist explains findings or a critic disputes them).

Journalists conducting interviews are sometimes searching for meat, sometimes for motion, and sometimes for both. When they are researching the facts (the meat), usually at the beginning of the reporting process, they may not need quotes or sound bites. Often they wait until they know the content of the story before they attend to getting quotes to add motion.

Scientists who complain that reporters have already made up their minds before conducting an interview may be right. But that doesn't mean journalists are uninterested in the facts. It may just be that they already have enough facts. If a journalist is getting a quote from a recognized expert to add gravity to her report, for example, she may already know what the scientist has said on related topics in the past, and can probably guess what the person might say on the topic at hand. A careful reporter has already surfed the Web, looked at earlier news reports, or talked to other people before beginning the interview. Since at this point she is in search of motion, not meat, she may already have made an outline of the story and expects to use a quote or sound bite to make a particular point. Her editor may even have instructed her to obtain a quote of a very specific nature, such as somebody who is excited about the research or doubts the research or questions the morality of the research. Has this journalist already decided on the angle of her story before going into the interview? Probably. Is she unconcerned about getting her facts straight? Probably not.

The Balancing Act As American journalism ascended out of its roots as the partisan mouthpieces of political parties, it needed rules of the road. As an alternative to the blatant bias being replaced, the new ideals chosen were objectivity, accuracy, truthfulness, all enlightenment ideals that made journalism almost a scientific endeavor. Over the years journalists have discovered that however worthy these goals it is difficult if not impossible to be certain when they have been attained. As a substitute, the profession has made "balance" and "fairness" standards of good reporting. Balance to a journalist means including opposing views on contentious issues. The standard of fairness requires that these views be accurately represented.

For reporters, this method of attaining, or at least approaching, truth has many advantages. For instance, if a harried general reporter with no previous experience in the topic is assigned to write about some new climate change research, he will not have time to familiarize himself with the science to assess whether he is accurately reporting on the subject. It is much easier, and less time-consuming, simply to "balance" the author of the report with someone who has doubts about it. Critics of this style of reporting say that such shortcuts may be doing more harm than good, however. Sometimes, as in the case of doubters who question whether humans are having an impact on Earth's climate, skeptics don't deserve the amount of "airtime" they get using the standard practice of balance. Journalists searching for alternatives to using balance suggest that journalism needs new tools for trying to attain the profession's ideals.

Doug Starr, the co-director of the Science Journalism program at the Boston University Department of Journalism teaches his students to incorporate what he calls "balance through depth." By this he means that they should give the readers the context to understand how to interpret a skeptic's views. For instance, it could be that someone who says, contrary to the vast majority of the world's climate researchers, that humans are not warming the planet will turn out to be a Galileo who sees what almost all other scientists have missed. But a journalist who quotes him, according to this approach, would want to say that only a tiny fraction of dissenters in science, perhaps one in a thousand or fewer, end up proving everyone else wrong. "The age of the stenographer journalist is over," says Professor Starr. "It has to be replaced with the analyst journalist." Such ideas are gaining currency among younger science journalists. But it is probably too early to write the obituary yet for the use of balance as a shortcut for achieving truth and accuracy.

WHAT'S ON THE TUBE

Pick your poison: television or print. Is television more superficial than print? Is there less regard for the truth on the nightly news than in the daily paper? These questions aren't as easy to answer as it might seem at first. But the process of seeking to answer them sheds valuable light on how journalists work and how to interpret what they do.

On August 6, 1996, NASA administrator Daniel S. Goldin startled the scientific community and dazzled the public with the announcement that a team of NASA and university scientists had "exciting, even compelling" evidence that life existed on Mars three billion years ago.[28] In a press conference the next day, President Clinton said that, if confirmed, the research "will surely be one of the most stunning insights into our universe that science has ever uncovered."[29] The team's remarkable findings came from a study of microscopic grains in a meteorite discovered a decade earlier in Antarctica and widely believed to have come from Mars.

In a paper published in *Science* ten days after the dramatic announcement, the researchers said three independent lines of inquiry all suggested that life once lived on Mars.[30] First, the team discovered polycyclic aromatic hydrocarbons (PAHs), simple organic compounds sometimes associated with life, in the four-pound rock. The PAHs were found deep inside the meteorite and in higher concentrations than previously seen in Antarctica, leading them to conclude that the compounds had been brought with the rock from Mars, not absorbed during the thousands of years it lay undisturbed in Antarctica. A more sensational claim was that the scientists had discovered minute fossils of microorganisms, smaller than 1/100 of the diameter of a human hair. Finally, the team reported finding tiny particles of iron oxide similar to those seen in certain bacteria on Earth. The day after Administrator Goldin's announcement, NASA held a press conference so members of the research team could discuss the findings. William Schopf, a biology professor from the University of California, Los Angeles, who was unaffiliated with the research, also gave a presentation. Quoting Carl Sagan ("extraordinary claims require extraordinary evidence"), he presented a careful critique to demonstrate that "additional work needs to be done before we can have firm confidence that this report is of life on Mars."[31]

Almost as soon as word of the NASA team's stunning announcement was out, John Kerridge's phone began ringing. Kerridge was the lead author of *An Exobiological Strategy for Mars Exploration*, a report completed the previous year to advise NASA on how to search for evidence of life, either past or present, on Mars. "The first couple of days was insane," recalls Kerridge, adding that if he took a bathroom break "there would be six more messages" when he returned. Calls to his home and office came in at all hours from around the world. His most "bizarre" experience: a request to join a live call-in show in Australia one evening while he was reading in bed. Obligingly, the scientist agreed, bemused to find himself "sitting in bed in pajamas talking to commuters in Australia." At a frantic pace for the first several days and then less frequently for the next several months, Kerridge told one journalist after another, "the conclusion is at best premature and more probably wrong," an assessment that is now accepted as correct by most knowledgeable scientists.[32] The PAHs, he said, could have a non-biological origin, as was presumed to be the case for such compounds when found in interplanetary dust particles collected by high-flying aircraft and balloons. The same could be said of the iron oxide particles. Finally, Kerridge noted, the fossil-like objects were much smaller than any fossils ever seen on Earth. They also appeared to be missing structures that paleontologists look for in such fossils. He therefore believed these objects were probably just suggestively shaped mineral deposits.

While print journalists seemed receptive to his critical assessment of the meteorite research, Kerridge concluded that television producers were uninterested in nuances that would have complicated their simple storylines. He was interviewed by both ABC and NBC, but didn't appear in the programs these networks aired. Appearances on CNN and ABC's *Nightline* were cancelled at the last minute. In a subsequent letter to *Science*, Kerridge described his experience and advised other scientists: "if you want to be on television, tell them what you think they want to hear. If you want the public to know the truth, stick to print and radio."[33]

Is television really that bad? Yes and no. An analysis of print and broadcast stories about the Mars meteorite shows that indeed television did include fewer skeptical voices than print stories. The Television News Archive of Vanderbilt University (which contains recordings and transcripts of *ABC Evening News*, *CBS Evening News*, *NBC Evening News*, and *CNN WorldView*) lists fifteen shows reporting on the meteorite research in August 1996. Only four included an interview with a scientist who had doubts about the evidence (Kerridge didn't appear in any of these stories). In contrast, about half of the 103 newspaper and wire service stories on the topic included in the LexisNexis database during August 1996 contained at least one quote from a scientist with doubts about the research (including eight stories quoting Kerridge).

It isn't possible to say why Kerridge failed to appear in the television reports. It could have been that he had a bad hair day or that his office was too noisy or that the producers didn't feel he made his argument cogently or, as Kerridge surmises, that producers considered his views irrelevant for their upbeat stories. (In the case of ABC's coverage, Ned Potter, who reported on the meteorite story for ABC News, says that Kerridge "landed on the cutting room floor" because "other people spoke more clearly, making fundamentally the same point he did.") One thing that is certain is that television generally prefers simpler, more one-sided stories. There is also a good reason why journalists sometimes don't appear to be using interviews to learn what the interviewee has to say (but, nevertheless, are not wasting anyone's time).

Network and local television news stories are almost always brief, a fact that reporter Ned Potter admits makes them "less detailed." In one study of 154 local stations by the Project for Excellence in Journalism, 42 percent of all news stories were less than 30 seconds long.[34] Only 31 percent were more than 60 seconds long. The *sports* segment of an average half-hour local evening news show was twice as long as just about any news item ever run on local news. A tiny fraction (7 percent) of local news stories analyzed by the group were worthy of being described as "high-level enterprise," due to some initiative on the part of reporters beyond merely attending a news conference or appearing at the scene of a crime or accident. A related study of network news showed that half of all evening stories contain either only one source (interviewee on camera) or no sources at all (the figure is 62 percent for morning news stories).[35] But, responding to Kerridge's accusatory letter in *Science*, Potter says network news shows are no less skeptical than newspaper stories: "Not in my experience." Nonetheless, the case of the Mars meteorite suggests that in the process of cooking science subjects down to a couple or several minutes of television, some important ingredients are boiled off. As noted above, only about one-quarter of the television reports about the Mars meteorite in the Vanderbilt database had a skeptical voice. This is consistent with a growing tendency for television reports to contain only one voice other than the reporter, creating a product journalists call single source stories. In contrast, about the same fraction of the print stories featured two such skeptics, giving these stories more nuance. When especially contentious or complex topics are covered on television news, more voices are added, but only with great reluctance. This explains, in part, why television reports with complicated logic are a rarity.

As we've shown in some detail in this and previous chapters, much can go wrong when you take that call from a reporter, or place such a call yourself. Of course, most scientists are never publicly rebuked for sensational headlines about their research, like Lee Hannah, or forced to spend days without sleep doing damage control, like John Gordon. We have focused on such outrageous cases not to scare you away from contact with the press, but because such stories starkly illustrate that the scientific community needs to communicate better. The situation desperately needs to be improved. Understanding how the media works is the first step to remedying the problem. The next step is what the following chapters are all about.

CHAPTER 4

Do You Hear
What You're Saying?

IF REPORTERS DON'T understand what a scientist is saying, how can they translate it for their audience? Or, put another way, if scientists aren't clear and concise, how can they expect busy reporters to get the story right?

Here's the good news: you can do something about it. There are several considerations scientists need to take into account during their interactions with the press, but one stands out: you must prepare clear and concise messages to express the facts or views you want to get across. John Funk, a reporter at the *Cleveland Plain Dealer*, urges scientists to "break down the information, using analogy if necessary, to help me understand what it is they are doing." This is not something you can do on the fly once you're already on the phone with a reporter. You have to do it before you sit down for an interview or write a press release.

In the political arena, we often hear "messages" compared with "spin"—the purposeful shaping of information to obscure or deceive. That is not what we're talking about here. "Messages" in this context are just a way to focus what you want to say in a way that the reporter's audience will understand and remember.

Think about it this way: if reporters don't have to decipher what you say, or guess what your point is, they are much less likely to misquote you or inaccurately describe your research or positions. "Communicating in the media is different from a discussion with a colleague or a presentation at a professional conference," says Christine Jahnke, president of Positive Communications, which prepares political candidates and subject experts for nationally televised debates and media appearances. "Most media interviews are edited," Jahnke notes, "thus the spokesperson has less control over how the information is presented and must rely on the reporter's ability to understand an issue. It is critical that scientists learn how to package what they want to say so that it is not edited in an inaccurate fashion."

If you develop focused messages that are interesting and understandable in advance, you'll have greater control over the information the reporter uses, and you will be less likely to see your name attached to something you either didn't mean to say, wasn't relevant, or was misconstrued. No longer will you be on the defensive, answering every question that comes your way as if you're on the receiving end of a Serena Williams serve. Instead, you'll be steering the reporter to the story you want the public to read or hear. This is not about spinning a reporter or hiding inconvenient facts. This is about communicating your views or research in a form that ensures the final story is both accurate and compelling.

So what is your message? Conceptually, it's simple—your message should be the most important theme, fact, or opinion you want the public to know. An effective message is clear, concise, and relevant. Ask yourself, "How would I want the headline to read when I open tomorrow's paper?" The answer to that question is your main message. Two or three other secondary messages can provide themes for the rest of the story.

At first, it may seem difficult to narrow your views or research down to three or four main points, especially if you are trying to encapsulate a large study or your views on a complex policy-related matter. Still, some scientists do it all the time. Kai

M. A. Chan, an expert in conservation biology and environmental ethics at Stanford University, says focus is the key. "Figure out the take-home message, don't stray from it too much, and say it in a sufficient number of ways that they understand it," Chan advises. By focusing on only three or four points, you ensure that a reporter knows what is most important. "Have unambiguous answers ready that they can't change the meaning of," adds forest ecologist Lee Frelich.

A segment about your work on a local television news program usually won't last more than a minute and a half. In that time it would be impossible for a reporter to cover more than three or four main points. And while newspapers often devote a significant number of column inches to a story, print reporters generally limit the scope to three or four main points as well; they just cover each in greater depth. If, by contrast, you were to stress six or eight or more points, the reporter—not you—would have to determine the main message. That increases the odds that a story won't match your intent.

Therefore, develop three to four messages for every interview, regardless of whether it's with a newspaper, radio, or television reporter. For each message you will then need to develop talking points (submessages). Think of your main message as the headline of a news story, such as "Scientist Discovers Poisonous Tulips." The story's deck (the text between the headline and body of the article) might be your second message: "Finding explains illnesses in backyard gardeners." Another subhead (rare in newspaper stories) might introduce your third message: "Gardeners should wear protective gloves and masks." The facts within the body of the article that support these messages would be your talking points.

KEEP IT SIMPLE (BUT NOT SIMPLISTIC)

The reporter is a surrogate for the public. As you choose the messages you want to deliver to the media, keep in mind the actual target audience: the television viewer, newspaper or magazine reader, or radio listener. "Frame the issue clearly for the public, not the reporter," says Lawrence B. Cahoon, who specializes in biological oceanography and environmental science at the University of North Carolina, Wilmington. Your goal is not to show the reporter how smart you are, but to communicate in plain language so the average newspaper reader or television viewer can understand.

This point was stressed again and again in our interviews with scientists. Brett Taubman, an atmospheric scientist at Pennsylvania State University, says, "I have found it best to summarize the findings of our work into concise, understandable, interesting sound bites. The media will never publish the long, pedantic soliloquies."

And what will the "average" person understand? Most newspapers write for people who can read somewhere around the eighth-grade level, but for some the target is fifth grade and for others, high school.[1] The key to making your messages comprehensible is short sentences and short words. Or, as Erica Fleishman, a conservation biologist at Stanford University, puts it, try "to frame the issue in simple but not dumbed-down terms."

"Dumbing down" means simplifying your message so much that its essence or meaning is lost. In addition to potentially hurting your reputation or embarrassing you, dumbing down your research or views is not in the public or media's best interests either. The adage "simple, not simplistic" sums up the best approach. Paula LaRocque, the retired writing coach of the *Dallas Morning News*, says, "One hallmark of intellect is the ability to simplify, to make the complex easy to understand. Anyone can be unclear."[2] How do you simplify without going too far? Here are a few quick tips:

Remember Your Audience As noted above, your audience is not the reporter. Oceanographer Jed Fuhrman advises scientists to talk to reporters "as if they are students interested in the subject but needing lots of explanation (and no jargon)."

No one would advocate that you should talk to another adult as if they were a child. Imagine instead that you are talking to an adult family member who is far removed from your work. For example, if you're an astrophysicist at the University of North Carolina, imagine how you would explain your work to your uncle Ned in Illinois who runs a nail factory, or your sister Sally, who works as a financial advisor in Maine. Ned and Sally might be very intelligent, but neither one has a clue how astrophysics works. They are your audience. The *Cleveland Plain Dealer*'s Funk says don't assume "that I aced calculus and can run with the big boys *or* that I majored in physical education and have calluses on my knuckles. Most newspaper reporters start out as generalists with a lot of curiosity. We are not scientists, and if we were, we'd not likely be able to write an article that very many people would be able to read."

Avoid Technical or Scientific Jargon Christine Jahnke from Positive Communications says, "Scientists must translate their work so that the general public will understand it. This can be done by using examples and avoiding jargon. However, scientists will be most effective if they can talk about their research and findings in such a way that they show a connection with the day-to-day life of the average person (i.e., what is the impact of global warming on someone living in southern California). The ability to talk about science issues from the perspective of the audience member is essential."

Remember, most people in our country are fairly illiterate when it comes to science. When you must use a scientific term, explain what it means. Avoid abbreviations and acronyms. Keep your sentences as short as possible. Listen to how network news anchors speak—they report on complicated subjects with uncomplicated language.

Lee Frelich says he's learned from painful experience that if he doesn't get his message straight, the reporter certainly won't. In the late 1990s, for example, Frelich discovered that earthworms, which are not native to Minnesota but had been inadvertently introduced there by humans, were threatening native plants. He has given dozens of interviews on the topic, and first described his research with explanations such as, "We've found that after putting out a number of plots in areas with earthworms and without earthworms, the abundance of certain species of native plants is lower when you compare the areas with earthworms and areas without earthworms."

Frelich says the first ten or so interviews on the topic were "disastrous." Especially with television and radio news spots, the journalists managed to mangle or trim his quote so that he appeared to be saying the opposite of what he had really said. Frelich realized he had to "control the flow" of the interview. Now, unless he is talking about something with someone who really wants depth, Frelich comes up with just a few points ahead of time and sticks with them. "If they ask ten questions, I say the same thing in five or six ways." Now, because he has learned what journalists want, the articles make his points correctly. "I give them sound bites that are so cute they have to use them," he says. For example, he now begins his discussion of worms with the enigmatic statement, "One of the most widespread beliefs in society is that earthworms are good for the environment." About this quote, the media-savvy scientist says, "I challenge any reporter to botch that one. . .only an idiot wouldn't follow with another question." When the reporter predictably asks why earthworms would not be good for the environment, Frelich answers with this succinct message: "Many of our species of native wildflowers are going extinct because of these earthworms." Now, says the ecologist, "they make the points I want to make."

SOUND BITES THAT WORK

What Frelich's experience and others like it make clear is that scientists should work at making their key messages and talking points (submessages) suitable for sound bites. In our conversations with scientists, we've learned that the term "sound bite" has different meanings to different people, but in this context think of it as a way to communicate your message in a short, quotable form for print or broadcast media.

Long-winded statements rarely end up in newspaper articles and radio interviews, and are never used on television. Just like your main message, a good sound bite is a concise statement that frames an important point in a way that will be easily understood by the public. But a good sound bite also has a certain extra something—it is surprising, catchy, humorous, or sobering. "For scientists," says Dan Vergano, the longtime science reporter for *USA Today*, "the trick is to put yourself in the reporter's shoes—they need a summary quote that captures some of the field's interest in a topic, but delivers it forcefully. What I look for in a quote is (hopefully) a slightly off-kilter way of expressing the scientist's view of some topic." In other words, the quote is memorable. And that is especially important in today's news environment, where people are bombarded by information. If your audience can remember just one of your main messages because of how you worded it, you have been more effective than most in reaching the American public.

After analyzing hundreds of newspaper articles and radio and television interviews over the years, we've determined that the media prefer certain kinds of sound bites from scientists. To help you craft your own talking points and sound bites, we've developed a list of speaking styles that reporters look for and that scientists have used effectively. The overriding principles, as scientists with successful media experience made clear, are to avoid jargon, use active and colorful words, keep it short, and, most importantly, keep it focused on one of your main messages or talking points (submessages).

1. Put your message into perspective. Reporters look for quotes from scientists that summarize the impact of a research finding or policy decision and why it's important. Describe in short sentences what's at stake. Explain in easy-to-understand language what this discovery means for our understanding of people, the natural world, or the cosmos. "The simplest way to deliver one of these quotes," says Dan Vergano, "is to drop the 'is,' 'are,' 'were,' 'have been' verbs from your vocabulary and try an active verb instead, e.g., 'reveals,' 'spurs,' 'screws up completely,' and so on: 'The results completely screw up the way we usually think about fireflies.'" Vergano also suggests highlighting the implications for citizens, wildlife, businesses, or whoever or whatever might be affected.

Basically, you want to offer your main message in a quotable, sound bite-ready format. Browse through the quotes below and you'll see a common thread: the scientists capture the essence of their research or viewpoint succinctly and clearly, while explaining the implication (in other words, why the science is important).

> "This will really revolutionize our ability to collect high-precision environmental data."—Todd Dawson, University of California, Berkeley
>
> The *San Jose Mercury News*
> HEADLINE: Tiny Remote Sensors Could Reshape Research
> REPORTER: Glennda Chui
> August 12, 2003

> "These are really enormous geological events. Earth doesn't do this very often. Something about the size of these things shakes you to the very core."—Brian Atwater, U.S. Geological Survey and University of Washington

USA Today
HEADLINE: Clash Of Continents Unleashed Deadly Waves
REPORTER: Dan Vergano
December 28, 2004

"This is important information for policy makers charting the future course of the Earth."—Taro Takahashi, Lamont-Doherty Earth Observatory (Columbia University)

The *Seattle Times*
HEADLINE: Industrial-Age Carbon Dioxide Found in Oceans: Half the Man-Made Gas Absorbed There; Acidity of Water rising, could harm sea life
REPORTER: Sandi Doughton
July 16, 2004

"The way forward is clear. We (in California) need to take a leadership position on putting the world on course to lower emissions."—Christopher Field, Stanford University.

The *Sacramento Bee*
HEADLINE: Study: Major Changes from Warming
REPORTERS: Edie Lau, Stuart Leavenworth
August 17, 2004

2. Your message should come from the heart. Your passion about a subject should be obvious from your choice of words, not just from your body language or the inflection of your voice. Use vivid and vibrant words to describe your reaction to something. People want to know how it makes you feel—excited, surprised, disappointed, skeptical. Dan Vergano advises to "just honestly say what you think: 'Beats me if this will pan out, but if it does, we'll all be wearing t-shirts with these guys' names on them.' If that's what you think, why not say so?" The examples below succeeded in showing reporters (and the public) what the scientist truly felt.

"If I didn't have all of these facts in front of me, and you came up with a universe like that, I'd either ask what you've been smoking or tell you to stop telling fairy tales."—John Bahcall, Institute for Advanced Study (Princeton University)

U.S. News & World Report
HEADLINE: The gods must be crazy
REPORTER: Charles W. Petit
September 8, 2003

"It is always exciting when you find something that you absolutely don't understand."—Brian Marsden, Harvard-Smithsonian Center for Astrophysics

The *Boston Globe*
HEADLINE: Astronomers find small, cold world
REPORTER: Gareth Cook
March 16, 2004

"I almost fell out of my chair" upon seeing the first pictures.—Brian McNamara, Ohio University

Cleveland Plain Dealer
HEADLINE: Black hole blast stuns scientists
REPORTER: John Mangels
January 6, 2005

"It's impossible not to be excited, and I come from the skeptical side."—Clemens Burda, Case Western Reserve University

Cleveland Plain Dealer
HEADLINE: Tiny science expects to reap big advances: Nanotechnology summit starts Monday
REPORTER: John Mangels
October 24, 2004

As the next examples demonstrate, sound bites are especially effective if you can summarize your research or opinion on a subject while mixing in your personal feelings.

"We are looking to the future by looking at children. It's very frightening."—S. Jay Olshansky, University of Illinois at Chicago

Milwaukee Journal Sentinel
HEADLINE: Wave of childhood obesity could cut life expectancies: Researchers predict crush of maladies; critics say medical, lifestyle advances can mitigate problems
REPORTER: John Fauber
March 17, 2005

"I'm so stunned I don't know what to say. This is going to have a devastating impact on the space program."—Alexander McPherson, University of California, Irvine

Orange County Register
HEADLINE: Science community reels: Sadness overwhelms local engineers and scientists who have been part of space programs
REPORTERS: Gary Robbins, Pat Brennan
February 2, 2003

3. Paint a picture with your message. Because so many Americans don't understand the language of science or how science works, one of the best ways to help them understand is by painting a picture with words. Reporters like to personalize issues, so tell a story that illustrates your point. Or describe your research with vivid words, like the narrator of a novel. Quotes that use descriptive language to set a scene are a powerful and effective way to convey your messages.

"Pigeons bob their heads while they walk, which makes them look like morons, and so people assumed birds only have the moron part of the brain."—Tony Reiner, University of Tennessee

The *Washington Post*
HEADLINE: Bird Brains Get Some New Names, And New Respect
REPORTER: Rick Weiss
February 1, 2005

"You get blood everywhere, slime everywhere and green poop, and that's if the fish doesn't fly to pieces."—Duane Chapman, U.S. Geological Survey

Minneapolis Star Tribune
HEADLINE: Volts may fend off invading carp
REPORTERS: Tom Meersman, Mark Brunswick
October 1, 2003

Perhaps the most effective way to paint a picture with your message is to come up with an analogy, metaphor, or simile to which people can relate. In most cases, you are part of a very small group that understands and works on your issue. So, when comparing your research or views with something more commonplace, think realist, not abstract.

"We are seeing the first few raindrops of a coming storm."—Leonard Zon, International Society for Stem Cell Research

The *Boston Globe*
HEADLINE: Britain allows cloning of human cells for research
REPORTER: Gareth Cook
August 12, 2004

"The Labrador Sea is like Grand Central Station for the global ocean climate."—Peter Rhines, University of Washington

Seattle Post-Intelligencer
HEADLINE: Climate theories run hot and cold: UW scientists
look at arctic clues
REPORTER: Tom Paulson
April 16, 2004

"It's something like driving your car through a severe hail
storm, but a hundred times worse."—Thanasis Economou,
University of Chicago

Cleveland Plain Dealer
HEADLINE: 'Stardust' collects space dust samples from
comet's tail
REPORTER: John Mangels
January 7, 2004

4. Relate numbers and methodology simply. Journalists usually
report data in their own words rather than quoting scientists. Scientists' sound bites are instead used to put numbers into context.
Sometimes, though, reporters will include quotes with numbers
that represent a trend, or dates that suggest a timeline. So if you
want to be quoted about specific data or dates, round them off
and explain what they mean in simple terms. A reporter wouldn't
find a quote like this very useful: "Our findings show that a lateral
movement of 21.4 meters over 63.94 seconds would require only a
very small lateral acceleration of 0.0004 g." The following quotes,
however, show that scientists *can* get it right.

"Over 2 billion years of time, these rocks have repeatedly
gone through the ringer."—Nicolas Dauphas, University
of Chicago

The *Baltimore Sun*
HEADLINE:Beginning of life on earth may be written in stone
REPORTER: Dennis O'Brien
December 17, 2004

"Could we really have lost 900 million tons of krill without anyone noticing? I don't think so. You would expect to see most of the predators in decline, and that doesn't appear to be happening."—Steve Nicol, Australian Antarctic Division

Los Angeles Times
HEADLINE: Antarctic food chain in peril, study finds: Krill have declined by 80% since 1976, researchers say.
REPORTER: Usha Lee McFarling
November 4, 2004

"If anyone was waiting to find out whether Antarctica would respond quickly to climate warming, I think the answer is yes. We've seen 150 miles of coastline change drastically in just 15 years."—Ted Scambos, University of Colorado

Newsday
HEADLINE: Antarctic glaciers found thinning faster
REPORTER: Earl Lane
September 24, 2004

Another approach to sound bites would be to touch on the rigorousness of the research or methodology rather than specific numbers or dates.

"The numbers are crystal clear. The analysis is impeccable. There is no uncertainty about this."—Peter Brewer, Monterey Bay Aquarium Research Institute

The *San Jose Mercury News*
HEADLINE: Studies: Carbon dioxide poses danger to sea life; legacy of fossil-fuel activity; changes in ocean chemistry
REPORTER: Glennda Chui
July 16, 2004

"This was just elegant. It was pretty. It was irresistible."—Nicholas Katsanis, Johns Hopkins University

St. Louis Post-Dispatch
HEADLINE: WU biologist streamlines disease search: May
lead to new treatments
REPORTER: Tina Hesman
May 14, 2004

If your research must carry certain qualifications, make sure you state those clearly to the journalist, but understand that reporters are more likely to quote you if there is no gray area in your view or research.

"I'm 100 percent convinced that these structures were caused by microorganisms. We see the same kind of microscopic, tube-like structures in lavas formed in marine environments today."—Martin Fisk, Oregon State University

The *Atlanta Journal-Constitution*
HEADLINE: Life started small, worked fast: Rocks show very early rise of microbes
REPORTER: Mike Toner
April 23, 2004

5. Speak in the vernacular. Americans watch about 250 billion hours of television a year,[3] so it's no wonder politicians like to convey messages that include references to popular culture. If you feel comfortable crafting a message that plays off of something on television (or the movies) without undermining your message or credibility, then by all means do so. Reporters are very likely to use such sound bites. However, if you feel uncomfortable with this approach, try a cliché instead. Most writing instructors would probably tell you to avoid clichés, but reporters like Dan Vergano encourage scientists to use them when it will help convey a point. Clichés are well understood by the public, so they can be an excellent way to get your point across.

To wit:

> "We're already looking at a pretty dire situation in the Yakima Valley. We should look at this as sort of a warning shot across the bow."—Philip Mote, University of Washington
>
> *Seattle Post-Intelligencer*
> HEADLINE: Drought here may be worst since 1992: Scientists fear dry winter may be glimpse into Northwest's future
> REPORTER: Tom Paulson
> March 9, 2005

> "We're waiting to see whether this roller-coaster ride will continue. It's anyone's guess as to what will happen next."—Willie Scott, Cascades Volcano Observatory
>
> The *Oregonian*
> HEADLINE: Rumblings on volcano are leaving geologists puzzled
> REPORTER: Richard L. Hill
> September 28, 2004

> "You get a lot of bang for the buck."—John-Paul Clarke, Massachusetts Institute of Technology
>
> The *Boston Globe*
> HEADLINE: Team designs a quieter way to land jets
> REPORTER: Gareth Cook
> December 21, 2003

6. Employ the astute quip. The best sound bite of all is probably the astute quip. These often seem like off-the-cuff remarks, but the best communicators have these sound bites planned in advance. They are typically short and may be clever one-liners, insightful statements packed into a few words, or just a powerful declaration. They have a ring to them that reporters love and audiences remember.

"I'm becoming skeptical about my earlier skepticism."—
Lawrence A. Crum, University of Washington

The *New York Times*
HEADLINE: Tiny bubbles implode with the heat of a star
REPORTER: Kenneth Chang
March 15, 2005

"There are no white cliffs of Dover on Mars."—Joshua Band-
field, Arizona State University

Los Angeles Times
HEADLINE: Science file: Study disputes Mars ocean theory
REPORTER: Usha Lee McFarling
August 23, 2003

"To common sense, quantum mechanics is nonsensical."
—William D. Phillips, National Institute of Standards &
Technology

Business Week
HEADLINE: Putting The Weirdness To Work
REPORTER: John Carey
March 15, 2004

"We are digging in Einstein's wastebasket."—Robert P. Kir-
shner, Harvard University

The *Boston Globe*
HEADLINE: Top 10 scientific advances of 2003
REPORTER: Gareth Cook
December 23, 2003

"We haven't found the magic pixie dust."—Edwin Minkley,
Carnegie Mellon University

Pittsburgh Post-Gazette
HEADLINE: A very persistent problem: Scientists, engineers
tackle controlling or eliminating seemingly indestructible PCBs

REPORTER: Byron Spice
August 23, 2004

"I'm scrambling around on the floor to find where my jaw dropped."—Bernard Wood, George Washington University

The *Baltimore Sun*
HEADLINE: Fossils of tiny human dug up in Indonesia
REPORTER: Dennis O'Brien
October 28, 2004

"It'll be interesting to see what we can find out by setting another eight 'postdocs' loose on the problem."—F. Sherwood Rowland, University of California, Irvine

The *Orange County Register*
HEADLINE: UCI gets $1.2 million for climate-change research: Lands' End founder, concerned about global warming, to fund eight 'postdocs'
REPORTER: Gary Robbins
May 19, 2003

Some of the best quotes marry different types of sound bites. For example, the next quotes are clichés used as quips.

"That really is the $64,000 mystery."—Tony Plant, Pitt Center for Research in Reproductive Physiology

Pittsburgh Post-Gazette
HEADLINE: Researchers find gene that kicks off puberty's changes
REPORTER: Byron Spice
February 1, 2005

"We now know that T. Rex lived fast and died young."—Gregory Erickson, Florida State University

The *Oregonian*
HEADLINE: T. Rex grew rapidly, but never had to shop for new sneakers
REPORTER: Richard L. Hill
August 18, 2004

The next quote mixes a quip with a quote that paints a picture:

> "A person's tombstone is unlikely to read, 'Killed by an asteroid.' But there are a lot of these rocks whizzing around in space."—Clark Chapman, Southwest Research Institute

> The *Orange County Register*
> HEADLINE: Looking for trouble: Deflect or destroy? Scientists ponder how to ward off Earth-bound rocks
> REPORTER: Gary Robbins
> February 21, 2004

Or how about a quote that paints a picture, then uses a cliché as a quip?

> "Much of Titan is ice. We're just seeing the first snippets of information—the tip of the iceberg."—Torrance Johnson, Jet Propulsion Laboratory (California Institute of Technology)

> The *Orange County Register*
> HEADLINE: Titan images point to erosion
> REPORTER: Gary Robbins
> January 15, 2005

A WORD ON MODERATION

Our emphasis on sound bites doesn't mean that you need to string together a succession of witty analogies and quips. In the ideal interview, you'll pepper your conversation with sound bites that reflect your messages, then continue to use clear and concise language to describe other talking points (submessages) and details. Reporters have antennae for sound bites, and they know them when they hear them. That's why the quotes above ended up in print, even if that wasn't what the quoted scientist intended.

The bottom line is this: in all your interactions with the press—whether it's an interview with a general assignment reporter at your local newspaper or a top science journalist at one of the nation's leading media outlets—you should know beforehand what you want to say and how you are going to say it. That way, reporters will use the quotes you want them to use. They are just waiting for you to give them something good.

CHAPTER 5

Mastering the Interview

F OR TOO MANY scientists, being interviewed is a stressful experience. Even if you've prepared notes in advance and practiced what you plan to say, once the interview is underway you feel as though you are at the mercy of the reporter. You might wonder, "Is the reporter taking down my quotes properly? Does he or she understand the issue accurately? Why is the reporter asking tangential questions? Will my credibility and reputation be tarnished if he or she makes mistakes?"

Any of these scenarios can make you understandably nervous about how a story might turn out. Fortunately, it doesn't have to be that way. By taking control of an interview, not only will the interview experience be less stressful, but you'll be much happier with the resulting media story.

CHOOSE YOUR WORDS CAREFULLY

Reporters generally consider everything you say, whether in a formal interview or informal chat, to be "on the record" and usable in a news story. The problem with speaking "off the record" or "for background only" is that these terms mean different things to different media outlets. For example, consider the following excerpt from *Policy and Practice*, a publication of the American Public Health Services Association: "If you say, 'Can we go off the record?' and the reporter agrees, then you are party to an informal agreement that whatever you say during that period of time cannot be used in news copy."[1] Compare that with this excerpt from *Investment News*: "'Off the record' is a term used to refer to information given to a journalist where the identity of the source is protected. In this case, anonymity is requested before the interview starts."[2] So in one definition the information is off limits for use in a reporter's story; in the other it can be used, but without attribution. There is a similar confusion around "On background." The APHSA says, "Talking to a reporter on background means that they can use the information relayed in the conversation, but not attribute it to a source."[3] *Investment News* says, "'On background only' is when information is supplied to a journalist and can be used only to enhance the way a reporter views an issue."[4] The *Washington Post*, like many newspapers, urges its reporters to have their sources speak on the record. "If not, we prefer that a source be 'on background' so we can use the substance without identifying the source," says Leonard Downie, Jr., executive editor of the *Post*. "We avoid having sources go 'off the record,' because that means we could never use what they say, even to seek corroboration elsewhere."[5]

Given these various interpretations, our advice is for you to feel comfortable enough with everything you say to be able to stand by it and have it on the record. However, if the need arises for speaking off the record or on background only, work with the reporter to reach agreement on what these terms mean for your interview. Think carefully about the agreement before proceeding; if you are uncomfortable with any of its conditions, then it is best to decline the interview.

The only circumstance in which we would recommend talking to a reporter off the record is if you've developed a relationship of trust with the reporter over several years. In this case, the reporter values you as much as you value the reporter, and he or she is unlikely to do anything to tarnish that relationship. However, you should still ask for a clear explanation about the terms first.

DO YOUR HOMEWORK

Before you are interviewed, you should conduct a little research. Who is the reporter? What is his or her usual beat? What's the deadline? What's the thrust of the story? How can you be helpful? The extra minute it takes to ask these questions pays big dividends. "I always ask if the reporter is up against a deadline and how much space the story is likely to get," says Thomas Yuill, a specialist in epidemiology of animal diseases at the University of Wisconsin-Madison. "That encourages good, two-way communication and gives me a realistic idea of what they have to deal with on their end." Guy R. McPherson, an expert in conservation biology and terrestrial plant ecology at the University of Arizona, reiterates this point, saying that a key to success is "determining the goal of the story early during the interview (or occasionally before it begins), then conducting research or mentally preparing myself for the interview (this often includes writing a few 'catchy' phrases on a notepad)."

Christine Jahnke of Positive Communications says there are six key questions you should ask before every interview:

1. What is the topic of the story?
2. What is the reporter's angle?
3. When is his or her deadline?
4. If the interview is for television or radio, will it be live, live-on-tape, or edited? (These are described in more detail later in the chapter.)
5. When and where will the interview take place?
6. Who are the reporters' other sources?

The answers to these questions can inform how you proceed. For example, if you discover that a general assignment reporter with no history of science reporting is writing a story, you will have to exert extra control over the interview. Peter Spotts from the *Christian Science Monitor* offers this advice: "If you get someone who clearly has no background in the subject matter—and is too self-conscious or too proud to let you know that up front—be willing to step them through the issue at hand. You might ask them if they've 'read (your study)' yet. If they haven't, you can be sure they will require some extra handholding on the subject."

INTERVIEW WHEN YOU'RE READY

Taking the time to learn a reporter's needs and expertise, and preparing your talking points accordingly, is an important step toward a successful interview. There might be times, though, when you receive a call out of the blue from a reporter looking for information that isn't on the tip of your tongue. Ask the reporter if you can set up a time to conduct the interview and take a few minutes to prepare. If the reporter is on a deadline for today, then ask if you can talk in ten or fifteen minutes. Even reporters on immediate deadline can usually wait five minutes as long as they know you'll be calling back quickly.

Some scientists, such as Dr. Carol Sing, an environmental pollution control expert in Henderson, Nevada, have learned firsthand what can happen when you are interviewed before you're ready. "I engaged in an informal conversation with an environmental reporter from the *Las Vegas Review Journal* prior to the formal interview," says Sing. "To my chagrin, some of my informal comments were quoted out of context in the news article." Laurel Standley, an expert in river water quality with Watershed Solutions, LLC, in Beaverton, Oregon, had a similar problem. "My most recent experience with a reporter was being asked to comment without advance notice and not receiving guidance on what the reporter was most interested in," says Standley. "I bumbled with too many bits of information—in hindsight I would have asked exactly what she was interested in or just picked one issue to discuss."

STAY ON COURSE WITH A "MESSAGE COMPASS"

In the previous chapter, we explained how clear and concise messages can help make your views and research understandable to the general public, while quotable sound bites will help ensure the reporter uses the words you want. Consider these as points on a compass that will help you guide the interview forward. If you have four main messages, picture your main message at north, your second message at east, your third message at south, and your fourth message at west. Between each of these points are your submessages (talking points) and sound bites. In every interview try to only use the messages and talking points on the compass. This approach will help you reach your ultimate goal: a story that accurately reflects the most important aspects of your research.

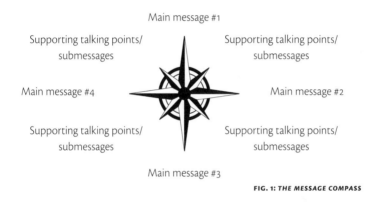

FIG. 1: THE MESSAGE COMPASS

Why a compass? The traditional setup of an interview, in which reporters ask questions and you supply answers, may well make you feel as though you are ad-libbing rather than in control of the story you want told. Using a message compass, on the other hand, permits you to go into an interview with a clear idea of how you want your messages to appear and allows you to proactively communicate these messages so the story is reported as accurately as possible. You always know what to say, no matter where the interview goes, because in every direction there is either a main talking point or a submessage. And unless your talking points are linear, you can switch between those messages when it makes sense.

REPEAT, REPEAT, REPEAT

If you have an important point to make, but only say it once during an interview, how will the reporter (or the reader, or the radio listener, or the television viewer) know it's important? You need to communicate your main messages as often as possible to make sure your story gets told. "Repetition is key to being heard," says Christine Jahnke.

STRAYING OFF COURSE

It is easy to stick to your talking points when reporters are asking you questions directly related to your messages. But what happens when reporters start asking questions far afield from those you want to discuss? If you start answering questions with responses not on your compass, the reporter will almost certainly take the story in a direction you don't want to go. To keep that from happening, answer the reporter's question as quickly as you can, and then go back to your main messages, which will direct the interview in the direction of the story you want told. This technique is called "bridging."

Certain phrases allow you to create a bridge from a topic off your message compass back to your main talking points. Jahnke suggests the phrase "and here's what I can tell you. . ." to get you back on your main messages. Other useful phrases are "but I do know that. . ." or "what's really important here, is. . ." For example, in response to a question on a topic beyond your expertise, you could say, "I haven't studied that area. Here's what I can tell you, though, about my research. . ." Or, "I don't know the answer to that, but what is really important when understanding this research is. . . "

THE INTERVIEW'S OVER; NOW WHAT?

It's important to remember that the interview is not over once the reporter stops asking questions. Mark Schleifstein, a staff writer who covers the environment for the *New Orleans Times-Picayune*, says, "I suggest that at the end of interviews, the scientists should ask questions to make sure the reporter understands the subject matter." Edie Lau, the science reporter at the *Sacramento Bee*, suggests taking it a step further. "If you are working with someone for the first time and aren't sure whether the person understands the subject to your satisfaction, don't be shy about asking them to get back to you before they finish the story. Most reporters will not submit an entire draft to you for review, but many are willing to go over passages, especially technical passages, to ensure accuracy." Some reporters, especially those writing for magazines, will offer you the opportunity to review copy, or even request that you do so.

Wildlife biologist Warren Aney suggests that scientists "offer to review anything the reporter writes for accuracy and completeness, without implying or exerting control over content (some reporters will welcome this, some will absolutely refuse to do so)." Dr. Kristen Sellgren, an astrophysicist at Ohio State University, adds, "It helps tremendously if I get a copy of the draft article to proof-read before the article goes to print. I can catch and correct some of the worst mistakes in describing the science."

Peter Spotts agrees that you should ask to review material, "but don't get grumpy if the writer says no." He adds that you should "never ask for prepublication review as a condition of participating in the interview. If you have doubts, decline the interview. Prepublication peer review may be part of science, but it's not part of journalism. Many publications forbid sharing stories with sources prior to publication."

If you are very nervous about conducting an interview over the phone, you can request that the reporter email you questions. This not only allows you some additional time to formulate a response, but also guarantees that your quotes are used properly.

ADDITIONAL ADVICE

Generally, you should approach every interview the same way, whether it's with a print, radio, or television reporter. In addition to the recommendations we've given thus far, keep the following tips in mind:

Keep it Short Brief answers are clear and easy to understand. They also make good sound bites. Christine Jahnke advises interviewees to keep "answers to concise statements of about 20 seconds," especially during television interviews.

Never Guess If you are unsure about a statistic or fact, tell the reporter you need to double-check the specifics and will get back to them. If you are being interviewed live, just tell the reporter that you don't know the answer, and bridge the conversation back to your message if possible. Alvin Saperstein, a physicist at Wayne State University, says the key to his interactions with the media is "openness, honesty (admit what you don't know), and completeness." Reporters will not think less of you if you need to check your research.

Be Yourself Just because you're going to appear on television doesn't mean you need to sound stilted, suave, or programmed. "Be pleasant and light," says Henry Pollack, an expert in geophysics and climate change at the University of Michigan, "not stern and ponderous."

Qualify When Necessary Tell a reporter directly when there are shades of gray on an issue. Richard P. McCulloh, a geologist with the Louisiana Geological Survey and Louisiana State University, suggests that you "qualify everything as you consider it appropriate, and continually reemphasize the qualifications as essential aspects or components of the technical content."

Don't Get Angry At some point you may find yourself on talk radio with antagonistic interviewers or experts with a different point of view. Whatever incorrect or inflammatory statements are made, don't lose your cool. Calmly rebut their points and communicate the messages you want the public to hear.

Be Friendly—But Stay Professional If a friend or colleague refers a reporter to you, you might feel tempted to let your guard down and be more candid than usual. That's fine, as long as you stick to your main messages. Don't speak off the cuff or say things you don't want in the media just because you have good rapport with a reporter. In the end, he or she is still after a story.

Don't Assume Brilliance (or Ignorance) Just because a reporter didn't get an advanced degree in science doesn't mean he or she is not smart. To make sure you're communicating your messages at a level that isn't above or below the reporter's understanding, ask if you're giving them what they need. "If the scientist has started to get the sense that the journalist isn't quite understanding, it's worth slowing down a little bit and exploring that," says Richard Harris, a science reporter for National Public Radio (NPR). "One thing that I do that's helpful—and a scientist might suggest this under certain circumstances—if I'm talking to someone and they're talking in jargon, I will paraphrase, and I'll say, 'Is this right? Is this a good way of putting it?' That gives me the opportunity to make sure I'm understanding it."

LIVE OR TAPE?

For radio and television interviews, it is important to know ahead of time whether it will be live, live-on-tape, or edited. A *live* interview is broadcast over the airwaves as you speak. There is no editing involved and everything you say will be heard by the radio station's listeners or seen on television. Most live interviews are only a few minutes long, so you need to be extra sure to communicate your main messages, and communicate them quickly—a four-to-five-minute interview seems much shorter when you are actually on the air. In longer live radio interviews, you have the opportunity to go into more detail with your submessages, but you should still try to stick to your message compass. Many listeners skip between stations and can tune out quickly, which makes repeating your main messages all the more important.

A *live-on-tape* interview is recorded as if it were live, but is aired at a future time (such as NPR's "Fresh Air"). You should approach the interview as if it were live, since the broadcasters want the piece to be seamless. Check with the reporter before the interview begins about whether you can redo part of the interview if you make an important mistake.

Finally, there is the *edited* interview. This is the most common interview format for television, and is used frequently for stories on NPR and other radio stations. Before the interview begins, ask how much of the interview the reporter plans to use. Most interviews are edited to twenty seconds or less, which translates into basically one sound bite. Given that, think about the main message you want to get across and keep repeating it. For longer edited segments, follow the same guidelines of a longer live interview. And, as Richard Harris suggests, "Don't be afraid to say, 'Oops, I slipped, let me start this over again.'"

RADIO INTERVIEWS

The old saying goes, "It's not what you say, but how you say it." For radio interviews, what you say *and* how you say it are both important. Listeners can't see your facial expressions or gestures as you present information, so it's important to talk expressively. A monotone voice can make even the most exciting research sound dull. In preparing for radio interviews, Aric Caplan of Caplan Communications recommends that scientists record themselves as they practice answering questions. "No better method exists to review and critique the rhythm and sound quality of someone's speech. . . . Hearing one's own voice can sharpen even the most experienced spokesperson's delivery."

Feel free to have your compass of main messages and talking points in front of you when you are conducting an interview over the phone, but don't read the points verbatim. "When people try to read what they have to say, that never sounds like they're talking," says Richard Harris. "It needs to be conversational. I like to have a conversation with someone, and during that conversation people will say interesting things that I pull out. It doesn't need to be a performance. Just listen to the questions and answer them." This is true for live, live-on-tape, and edited interviews

During the interview, Caplan suggests that in addition to asserting your key messages, highlight a website where listeners can get more information. He also stresses the importance of a "call for action." "After all," he says, "the purpose of speaking on the radio is to influence a particular outcome or desired result. The adage 'keep your audience wanting more' is true." If you are being interviewed for a radio show, ask questions in advance about its format. Are you representing one side of an issue to produce a debate, or is it a one-on-one interview? Will there be call-in questions from the public or will the host ask all the questions? Check the station's website if you have any concerns.

If you are pressed for time, accept radio interviews that will be replayed or aired live during morning or afternoon drive times. "Interviews during drive times reach the largest, most influential audiences available." Caplan says that Thursdays are the best day of the week to have your interview aired, because "most major radio promotions, major announcements, and contest winners are named on Thursdays." Weekend radio shows do not have as large an audience as weekday shows; however, if you are looking for practice, they are a great place to start.

TELEVISION INTERVIEWS

Sitting in front of a television camera can be more intimidating than talking on the phone with a print or radio reporter. That is all the more reason to prepare your messages in advance so you can feel comfortable and confident about the direction of the interview. And, again, that means asking questions up front. "Know the parameters for individual interviews," advises Caplan. "Though it's unrealistic to anticipate all the questions, ask the producer ahead of time for the scope or direction of the talk. If they are not being clear, reconsider the value of your time."

A television interview should incorporate the same elements of a radio interview: three or four main messages, including a "call to action," and a website for viewers to learn more.

Thomas Yuill from the University of Wisconsin-Madison says, "TV stories are particularly challenging as they usually want just a few seconds, making it impossible to give them a complete story and sometimes compromising accuracy." That's what happened to climatologist/meteorologist Steve LaDochy. "Several years ago, I was interviewed on CBC-TV on global cooling and 'weird weather,'" he says. "After going over the many theories involved in climatic change and the important variables, my half-hour of comments were edited to one sentence: that we just don't know what will happen for sure. And the next twenty minutes of the show was devoted to an interview with a farmer in the north woods who said the winter was going to be bad because the muskrat fur was longer that year."

When you do a television interview in person (rather than via satellite), look at the reporter, not at the camera. "Don't worry about camera crew, don't worry about future television audiences," advises Morrow Cater, a former reporter and producer for Frontline, who now runs her own communications agency, Cater Communications. "You are having a conversation with one person—remember that and keep your attention on that personal connection."

In most instances you will be seated and the camera will set up to tape you somewhat off center, though sometimes it might make more sense to be taped standing, especially if you are outside or using a prop. For seated interviews, Cater advises, "Don't let your jacket bunch up around your neck. Sit on the tail, sit up straight, and look at the interviewer not the television camera." Aric Caplan encourages interviewees to position themselves in front of an appropriate graphic or backdrop if possible; your university or company might have a banner with a logo or website that you can use. Sometimes a camera crew might suggest an interesting "visual backdrop," such as a lab, for your interview. If the backdrop is not appropriate for your expertise or makes you uncomfortable, simply explain why their suggestion doesn't make sense and offer a solution.

FASHION FORWARD

Any clothing or accessory that might distract the viewer from your message should be left at home. If you are being interviewed in the field or in a laboratory, wear clothes appropriate for that setting. For office or TV studio interviews, wear professional clothes that fit well and are comfortable. Christine Jahnke suggests the following wardrobe choices:

Women

- Business suit, dress, or pantsuit
- Fabrics with texture: wool, linen, cotton
- Rich colors: turquoise, royal purple, red (choose flattering shades)
- Cream- or pastel-colored blouse
- Makeup with a powder base; matte lipstick
- Jewelry with a dull (not shiny) finish, such as pearls
- Eyeglasses with rimless or light-colored frames and non-reflective lenses

Men

- Gray or navy blue business suit or blazer
- Off-white, light blue, or gray shirt with long sleeves

- Tie with a simple pattern
- Face powder to prevent shine
- Long socks to cover calves when seated
- Eyeglasses with rimless or tortoise shell frames and non-reflective lenses

What Everyone Should Avoid

- Black, white, and shiny fabrics
- Busy patterns such as paisley, stripes, plaid, or florals
- Large, dangling, or flashy jewelry
- Bright gold or silver jewelry that will reflect lights

IF A STORY COMES OUT BADLY

Despite your best efforts, sometimes a reporter might accidentally misquote you or inaccurately portray key information. If this happens, you should let the reporter know right away so the media outlet can issue a correction and the reporter gets feedback so he or she won't make the mistake again. Mark Schleifstein encourages scientists to "request a correction, and if they're not satisfied with the reporter's response, they should go up the chain of command until they are, all the way to the publisher of the newspaper if necessary. Why not? Corporations do that all the time."

Always be courteous and professional as you raise your complaints. "Be understanding that reporters have unrealistic deadline pressures that may result in errors in their stories," says Schleifstein. If you are unhappy about a headline, remember that reporters often don't write (or, until publication, possibly ever see) the headline. Likewise, magazine cover stories or other large features may have an introduction (presented prior to the article) that was written by an editor. Understanding these distinctions can help you effectively approach a reporter with your concerns and get them resolved quickly.

CHAPTER 6

A Reporter's Most Trusted Source: You

A REPORTER'S JOB IS to find news and tell the public about it in print, over the radio, or on television. They don't always discover these newsworthy events on their own. Reporters have "sources"—people who keep them informed so they can do their job. Becoming a source should be one of your top priorities in working with the media.

To be a source for a reporter is another way of saying that you have developed a working relationship with a reporter. Stuart Pimm, the Doris Duke Chair of Conservation Ecology at Duke University, says that if scientists build long-term relationships with reporters, reporters will "know they can count on you for a story and, if not, that you'll provide them names of people who can."

Indeed, there are many benefits, both to you and to reporters, of becoming a source:

1. Keeping reporters up to date. By developing a working relationship with a reporter, you create an open channel to update journalists on important scientific issues. This might include drawing attention to important research the reporter was unaware of, providing context about research and policy developments the reporter might have misunderstood, or providing updates about your research and the work of other scientists. Your ongoing contact with reporters will go a long way toward ensuring they have all the facts they need to do their job well. The lines of communication are likely to remain open as long as both parties continue to respect each other's constraints, such as deadlines and workload.

2. Becoming an on-call expert. Once you are a source, the reporter will call you for quotes, comments, or background information on stories related to your areas of expertise. You might also be called to help with fact checking. Stuart Flashman, who worked in biochemistry and molecular biology for Michigan State University, North Carolina State University, and the Stauffer Chemical Company, says that, as a source, "when an issue comes up, you're one of the people they think of to call and ask for comment."

3. Raising the visibility of your research or views. Put yourself in a reporter's shoes. If your job is to look for interesting stories and credible experts to quote, wouldn't you be more likely to seek comments from a scientist you know and trust, who keeps you informed? Scientists who aren't in contact with reporters usually don't get called out of the blue because the reporters don't know they have something to say and can say it well. By developing a working relationship with reporters you will have more access to them. Your ideas and views will be considered, and that alone will increase the odds of getting in the news.

4. Balancing coverage. Reporters have rolodexes just like you, and in those rolodexes are the names and phone numbers of sources and experts they call when writing their stories. But how do you know that those rolodexes aren't filled with people who don't share your views? Or even if they do share your views, isn't your perspective equally important? Your working relationship with reporters can help steer their coverage. You will be there when the writing starts rather than just reading it after the fact in the morning paper.

Reporters might contact you when a study is released or a new public policy is announced that could be of concern to you. Based on your recommendations, the reporter might change the angle of the story or drop it altogether. "If I've established a good relationship," says Flashman, "the reporter will call back to ask for my response after having talked to a representative of the opposing viewpoint. I also think it helps to tell the reporter what you expect the opposing side is likely to say (to the extent you know it) and give your responses ahead of time. This 'preloads' the reporter with pointed questions for when they do contact the other side."

5. Gleaning valuable information. While scientists are sources to reporters, sometimes the roles are reversed. For example, suppose an organization notorious for its shoddy methods is releasing a report tomorrow that misuses some of your research. If the reporter has an advance copy, he or she might call you for comment before the report is made public. This intelligence can inform the work you do with other media outlets. Or, if the report alarms you, you can enlist the support of other colleagues to have responses ready when the report is released.

6. Opening other doors. If you have developed a solid working relationship with a reporter, he or she can help you navigate other areas of the media. For example, if you live in North Carolina's Research Triangle and develop a relationship with a reporter at the *Raleigh News-Observer*, the reporter might be able to suggest an editorial writer at the paper with whom you could meet. Or, if you want to write an op-ed, perhaps the reporter can recommend you to the op-ed editor.

As a scientist, you have an inside track on becoming a source for reporters. You have credibility as someone who "discovers things" and understands things others don't. Your credentials and training immediately position you as an expert commentator on issues related to your work. You can inform reporters about new research developments at your university, or on research taking place elsewhere in the country. You can critique policy developments related to your expertise at the local, regional, or national level. You can even be an advocate for science issues more generally. But you can only do this if reporters know you exist and know how you can be helpful to them.

HOW TO BECOME A SOURCE

It might surprise you, but most reporters welcome calls from scientists with something interesting to say. Mark Schleifstein at the *New Orleans Times-Picayune* says, "I urge scientists to contact me whenever they have something they believe is newsworthy or controversial—and even when they don't."

Stuart Flashman notes, "I've found that the best technique is to initiate the contact with the reporter, spend enough time (either in person or on the phone) to clearly explain the issue, make sure you've allowed the reporter plenty of opportunity to ask clarifying questions, and leave the door wide open for them to call back for clarification or follow-up."

Once you have figured out what kind of stories a reporter is interested in, you can be on his or her scout team for news. Reporters meet with experts and sources all the time, so you are not embarking on any venture out of the ordinary. Andrew Revkin, a science reporter at the *New York Times*, says, "The more scientists and journalists talk outside the pressures of a daily news deadline, the more likely it is that the public—through the media—will appreciate what science can and cannot offer to the debate over difficult questions about how to invest scarce resources or change personal behaviors."[1]

Developing relationships with reporters is like any relationship—they are built on mutual respect and understanding, common interests, and trust. They also ripen over time. Although some journalists and scientists are buddies, developing a relationship with a reporter is usually not a matter of making a new best friend. Nor is it a détente between two people on opposite sides of a battle line. You should strive for an amicable, professional relationship, just as you have with colleagues in other departments or at different institutions. Don't forget, though, that no matter how friendly your interaction, whatever you say might appear in print or on the air unless you explicitly make other arrangements. (See chapter 5 for more about speaking on and off the record.)

The bottom line is that many scientists never take it upon themselves to contact reporters, and when they do, it is only when they have research results to release—and even then, it is most often through a press release or a staff member in the university's media relations department. This is one of the biggest media-related mistakes scientists make. "I hardly ever get calls from scientists," says David Kohn, a science reporter at the *Baltimore Sun*. You should make developing and maintaining a working relationship with at least one reporter part of your ongoing professional work. "If a scientist has a great story idea, call me up," says Kohn. "We're just people. It's much more annoying to get calls from PR people who are paid to push stories and may not understand the issue. I'd much rather talk to scientists who are excited about their research."

Retired Oregon state wildlife biologist Warren Aney stresses the importance of becoming a source. "We tried to be proactive with the local press—informing them when we had something that we thought would be newsworthy. As a result, we often got positive and constructive coverage of these events. The press even would call us before publishing an article or letter that could be considered critical to make sure the facts were correct—they even would end up not publishing letters if we could point out important factual errors."

Target a Reporter, Not "The Media" In the United States, there are 2,370 daily and Sunday newspapers, 1,748 television stations, 13,525 radio stations, and many thousands of other specialty magazines and independent news websites.[2] No scientist has the time to reach even a tiny fraction of these media outlets. And even if one did, many of those media outlets would not be interested in what the scientist had to say. They have diverse audiences and therefore different demands for news. That's why you should think of the media not as the huge, amorphous industry, but rather as individuals trying their best to do their job. If you break it down that way, you can start to identify specific reporters who might be interested in your work.

The local level is the easiest place to have an impact with the press, especially if you have never worked with reporters before. "The environmental reporter for our local newspaper is enthusiastic, intelligent, and receptive," says David Strayer, a specialist in fresh water ecology at the Institute of Ecosystem Studies in Millbrook, New Jersey. "[The reporter] often contacts us for information (and has a good idea what sorts of information we can provide), and responds to our suggestions for news stories. Consequently, our research is often and accurately included in the newspaper."

With the exception of a few national media outlets, such as the *New York Times*, reporters will quote local experts when given the choice. David Kohn says, "I consider myself a national reporter—but my bosses do want us to be covering the local universities." The reasons are simple. First, local experts know the most about local conditions. Second, a local media outlet generally assumes it axiomatic that their audience is more interested in local voices. If you teach at the University of Washington, for example, a reporter at the *Seattle Times* is much more likely to be interested in your work than a reporter at the *St. Petersburg Times* in Florida. Likewise, a professor at the University of Texas will have much greater luck becoming a source for a reporter at the *Austin American-Statesmen* than a television producer in Providence, Rhode Island.

Unless there is an obvious choice in your local radio or television station, start by contacting one reporter at your local newspaper. Not only are newspapers more likely to cover your issues, often the stories that are discussed on radio and television first appear in print. If scientific issues are being completely ignored or only appearing through national wires such as the Associated Press, you should look for the reporter who covers the next closest issue (for instance, there might be a reporter who covers other developments at your university, or a general assignment reporter who wrote an article last month about a local environmental or health-related issue).

Monitor Coverage It might seem obvious, but to be in the media, you have to consume it yourself. If you don't already read your local newspaper, listen to local news radio, or watch local television news programs, do it. We're not suggesting adding another hour of work to your already busy life. Just peruse your local newspaper for a few minutes each day, and sit through a few television segments a month. The same goes for talk radio, where many Americans receive their news these days. As you anticipate being part of the news (or the process of making the news), pay attention to the details you might have previously ignored. Who are the reporters? Who is speaking on air? What is the reporter's angle on the story?

Keep track of which reporters cover issues most closely related to your specialties. For example, the *Newark Star-Ledger* has a science writer, while the *Omaha World-Herald* has a writer covering both science and environmental issues. At smaller papers, there might be a rotating group of general assignment reporters who cover a different issue every day. Of course, in some cases, local media outlets might be excluding science coverage altogether, which is all the more reason to get to know a reporter at the newspaper or station.

Set Up a Meeting Your initial interactions with reporters are solely to lay the groundwork for future collaboration. The goal of your first contact is, not surprisingly, to introduce yourself and ask to meet. In this first conversation you should briefly explain your area of expertise, then suggest another time to talk so you can share more about your work and learn more about the reporter's beat and what issues interest the reporter most. Most newspapers have websites where you can find the phone number for the news desk. If not, look in the newspaper itself or the phone book and call the main number. Gareth Cook, a science writer at the *Boston Globe*, says, "In general, the best time to reach reporters is in the morning, when they are less likely to be under intense time pressure." Even then, you never know if a reporter will be in the middle of researching a story. If a person at the main desk answers, ask to be transferred directly to the reporter. If you reach an answering machine, here's a sample message you might consider leaving:

"Hello, this is Dr. Sarah Littlefield at Case Western. I noticed that you've been covering science issues for the *Cleveland Plain Dealer* and I thought it would be useful for me to get in touch. I've been conducting research on fish populations in Lake Erie and wanted to fill you in on my work. I'd like to set up a time to talk or meet. I'm also interested in learning more about what kinds of stories you're most interested in. I can be reached at...." A message like this tells the reporter that you have something useful to share and are genuinely interested in learning how you can be helpful to him or her.

If the reporter does answer the phone, your script is much the same, except at the beginning, you should check to make sure the person isn't in the middle of work on a story or on deadline. Here, the scientist might begin: "Hello, this is Dr. Sarah Littlefield at Case Western. Do you have a minute?" If the reporter has time, Dr. Littlefield could follow the script much as above and set up a time to talk later. If the reporter is busy, just ask for a good time to call back.

Email is not as personal as an encounter over the phone, making it less ideal for developing a relationship. However, if you feel more comfortable initially reaching out in this way, it's better than not being in contact at all. Some newspapers will print a reporter's email address at the end of a story, but most don't, so you'll probably have to call the main number for the newspaper and either ask the front desk person or someone on the news desk for the reporter's email address. Usually folks are happy to share this information over the phone. If not, then ask to be transferred to the reporter's voice mail. The benefit of an initial contact email is that the reporter can read the message during a free moment, so you know you won't be calling when he or she is in the middle of researching or writing a story. The downside is that reporters' email boxes can be swamped with messages, so yours could be ignored or overlooked.

Get Some Face Time Meeting face to face is the best way to get to know a reporter. Out of the office, the reporter is less likely to be in a rush. The atmosphere will be more relaxed and might lead to a longer, more interesting conversation. You both will have a chance to put a face to a name, and maybe even learn a few things about each other not related to work.

Lunch is one of the best times to meet a reporter because it fits well with the rhythm of a newspaper reporter's day. Breakfast meetings are another option, though for some reporters a breakfast meeting isn't convenient, especially if they worked late the night before. Afternoon deadlines generally make dinner out of the question. If you can, try to meet reporters near their office. The less time they have to travel, the more time they have to spend with you.

If you meet a reporter for lunch, offer to pick up the tab since it was your suggestion. The reporter might accept, but more likely he or she will insist on paying or splitting the tab. There is a propriety and ethical standard at stake for some reporters. They don't want the appearance of doing favors in return for money (even if their meal costs very little). If the reporter does offer to pay, don't put up a fuss; just say thank you.

If you feel uncomfortable meeting a reporter for lunch (and you really shouldn't, since reporters meet sources for lunch all the time and might well take that time away from their desks anyway), another option is to invite the reporter to your office. This is especially useful if it allows you to do a little "show and tell." Is there something in the lab that brings your work to life or makes it more understandable? Are there are other scientists in your department that you'd like the reporter to meet?

Consider alternative locations if it would better enable you to communicate your messages and connect images to your words. Fred Rhoades, an expert in mycology and lichenology at Western Washington University and owner of a biological consulting business in Bellingham, Washington, shares his expertise with an environmental and recreation reporter of his local paper during hiking trips. The reporter writes a regular column called "Wild Things," featuring local organisms. "I feel she is providing a very real educational service to the paper's readers," he says. "Since one of my goals in life has been to foster the increase in knowledge about the 'little organisms,' I wholeheartedly support her in her projects." Chance meetings are also a good time to start building relationships. Michael Bretz, a specialist in experimental condensed matter physics at the University of Michigan, said he developed a working relationship with a *Science* magazine writer at a large conference. "We got to know each other there, so later telephone conversations were low key and productive."

Most reporters will take you up on any of the invitations

above—such encounters are part of their job. A reporter who declines is most likely facing an important deadline and will usually be willing to meet you at a later time. It is possible, however, that a reporter might suggest meeting at his or her office. If it's convenient for you, do it. The meeting will probably be more rushed than if you met somewhere else, but it might be your best shot to begin a relationship with this reporter. It also gives you the opportunity to see the reporter's work environment. (You might even ask for a brief tour of the newspaper's offices.)

If schedules or distance don't allow for a meeting in person, then by all means schedule a time to have a longer conversation over the phone. The downside to a phone meeting is that the reporter is still tied to his or her desk and might be distracted by other phone calls, emails, hubbub in the newsroom, or even an article on the computer screen. The whole interaction will be more perfunctory. Still, it's better than not contacting the reporter at all.

Prepare Your Messages Once an initial meeting is arranged, prepare just like you would for an interview (more on interview techniques in chapter 5). Think ahead of time about your main messages. These are the details about your work that you want to be sure the reporter remembers once the meeting is over. Focus on just three or four main points—you don't need to cover the whole landscape. Remember the goal here is to create a long-term relationship, so there will be opportunities for follow-up. You want to demonstrate to the reporter that your work is interesting and important—or that your views on policy matters are relevant—and that you would be an excellent resource.

At the meeting, be relaxed yet professional. Consider everything you say to be on the record, even if the reporter isn't taking notes (they may well remember a point and ask you about it later). Here are a few other tips we encourage you to follow during your initial meeting with journalists:

1. Take the lead. After introductions, jump right in about why you asked to meet. You might start by saying, "I wanted to fill you in on some interesting projects that we're working on, but also wanted to learn more about what kind of stories you and your editors are looking for." Then start talking. Remember, you suggested the meeting, so you should feel comfortable leading it.

2. Don't filibuster. Taking the lead doesn't give you license to monopolize the conversation. If you love your work and are excited about it, you might have a tendency to go on for a while. This could backfire, because your messages could get watered down or even ignored. Make sure the reporter has a chance to talk, too.

3. Ask the reporter questions. Many reporters won't open up about their own work without a little prodding. But, like you, most reporters enjoy their jobs, so if you can get them started, they might share interesting facts about their work, articles, editors, deadlines, and so forth. This information is valuable, because it can help you better understand how to present your research and views in a way that will be useful. Some questions you might want to ask the reporter during your meeting are: How broad is your beat? What stories have you enjoyed covering recently? What story angles are you looking for? What is your reading audience like?

4. Let the reporter know how you can be useful. Your main talking points should give the reporter a sense of your work and expertise. But if appropriate, let the reporter know you have contacts in other relevant disciplines within the scientific community, or are tracking national or local science-related policy. If the reporter is ever looking for an expert to make sense of an issue or deliver a quote, you will be in the position to help.

5. Remember to talk simply. As we have emphasized in previous chapters, it's important to communicate your messages using language the reporter (and the reporter's audience) will understand. Again, we are not suggesting you dumb down your messages or be condescending; just be clear and concise. The reporter knows you are smart, so you should feel no pressure to impress with arcane scientific information. If he or she can't understand you or thinks your language is too technical for the general public, you might have cooked your own goose. One exception to this general rule is when you are talking to a reporter who has advanced training or has developed a specialty in your field. Ask him or her to tell you how much detail you should go into when presenting your information.

INFORMATION VS. NEWS

When talking to reporters, keep in mind there is a big difference between information and news. Information is background and context that sets the stage for a story or fills in the gaps. News, on the other hand, is new information that is relevant or interesting to a significant portion of a media outlet's audience. News is generally what media outlets want to cover.

Before you contact a reporter, be it an initial contact or a regular follow up, determine whether you have information or news to pass on. If it is news, you want to communicate it in a way that will convey its timeliness and importance. By telling the reporter you have news to share, you are asking the reporter to put aside what she or he is doing. On the other hand, if you have background information or other general information to share about your work, you want to be make it clear that your information has no immediate news value and therefore is not urgent.

What is News? How do you know whether what you have to say is newsworthy, especially if it's not currently receiving media attention? Monitoring local media outlets will help in this regard (noting that what is newsworthy for one outlet might not be for another). You should also ask yourself the following questions to help decide if you have something newsworthy, or whether you can make something newsworthy.

1. *Is it happening now?* If the answer is yes, then the story can be considered "breaking news." (You should also be able to answer "yes" to one or more of the questions below.) Breaking news means that the story is extremely timely or urgent. If you are aware of breaking news, you should call the reporter you cultivated immediately. Here's the introduction of a breaking news story from the Associated Press:

> "Usually by now the Columbia River's spring chinook salmon are heading upstream over fish ladders in the tens of thousands to spawn. But not this year.
>
> 'It's a never-before-seen scarcity,' said Charles Hudson of the Columbia River Inter-Tribal Fish Commission. 'We're way behind, even compared to the historically low years of 1994-1995.'"[3]

2. *Is it a new discovery?* Reporters like to report on new research or new discoveries. Peter Spotts of the *Christian Science Monitor* looks for "research-result stories" involving "studies that help advance a field in significant ways or that represent an intriguing twist to a 'conventional wisdom' research problem." *Science, Nature,* and myriad other science-related journals are the source of a large number of such news stories, as are universities and other institutions. The beginning of this *San Francisco Chronicle* article is a good example:

> "With surprising and mysterious regularity, life on Earth has flourished and vanished in cycles of mass extinction every 62 million years, say two UC Berkeley scientists who discovered the pattern after a painstaking computer study of fossil records going back for more than 500 million years."[4]

Keep in mind that this article wasn't written *just* because it was a new discovery; It was also surprising.

3. Does it affect people's lives? What does your research mean to your average Jill or Joe? Will it have an impact on her or his quality of life? Will it have an impact on local business? Reporters usually look for news stories that resonate with people's lives; issues relating to children are particularly resonant. For example, here is the lead paragraph of a story on a children's health issue that appeared in California's *Contra Costa Times*:

> "The first study in the nation to link children's respiratory problems with traffic pollution found a 7 percent higher rate of asthma and bronchitis in children attending schools near busy roads and freeways."[5]

And one from the *Washington Post:*

> "Ever wonder why children can't hit the softball or tennis ball even when you throw it oh so slowly?
>
> It is easy to understand, vision scientist Terri Lewis said. 'Kids can't judge slow speeds. They can't tell that the ball's moving at all.' The solution: 'Give it a little more oomph. They'll do better.'"[6]

4. Does it affect places people care about? Although a story is more likely to be considered newsworthy if there is a direct impact on people or public policy, there are science stories that feature places or parts of nature—oceans, wildlife, the galaxy—that people care about or find interesting. The beginning of a story from Florida's *Bradenton Herald* illustrates this:

> "The coral reefs of South Florida will continue to decline, becoming little more than 'rubble, seaweed and slime,'

unless the government takes stronger steps to protect them."[7]

5. Is there conflict? Reporters love stories that are controversial or show evidence that conventional wisdom is wrong. That's why some reporters still cover the issue of global warming as if the majority of climate scientists weren't in general agreement about the problem. All it takes is one scientist to challenge the consensus viewpoint to make a story newsworthy. But conflict doesn't need to be just between scientists. Perhaps your research is about an endangered species that landowners and environmentalists are arguing over. Or perhaps a community is at odds with the local government over a proposed new policy. Here's an example from the *Daily News* in Jacksonville, North Carolina:

> "Some environmental activists and fisheries regulators aren't sure they agree with a petition that claims it's time to list the native eastern oyster as a threatened or endangered species. While it may be a good way to draw attention to the plight of the population, the oyster probably doesn't meet the qualifications, said Todd Miller, executive director of the North Carolina Coastal Federation. 'I don't think we're quite at that condition yet—particularly in North Carolina,' Miller said."[8]

6. Is it unusual? Reporters love stories out of the norm. These story angles might be curious, or funny, or even shocking. Consider the following lead from a *Seattle Times* article on our ancient ancestors:

> "The Neandertal waistline keeps growing and growing.
>   Our human cousins—a species that scientists believe died out about 30,000 years ago—make modern waists look

wasp-like, according to anthropologist Gary Sawyer, chief technician at the American Museum of Natural History in New York, and Blaine Maley, a former sculptor who is now a Washington University doctoral student in anthropology."[9]

7. Does it relate to something else in the news? "Why should I write about this now?" is a common question reporters ask when considering a request to cover a science story. Don't let this deter you. They just want to "hang" your information on a "news peg" to make it relevant to readers. For example, if you released a report a year ago, the reporter might feel that it's not new enough. But if your local town council is discussing a new measure that is related to your work, the reporter might well include your information as part of the story. Or a tragedy happening elsewhere in the world might give a reporter a hook to write a story about the likelihood of it happening closer to home. The lead paragraph of this article from the *Chicago Tribune* is a good example:

"The East Coast of the United States, with no advance warning system or equipment to detect a giant wave in the Atlantic Ocean, remains extremely vulnerable to tsunamis like the one that devastated Southeast Asia on Dec. 26, a group of scientists warned Tuesday."[10]

8. Is it connected to someone or something important? Famous people and places are always in the news. It's why you can't go through the check-out line at the grocery store without seeing a cover story about Michael Jackson or Jennifer Lopez. There aren't many celebrity scientists (at least not in popular culture), but perhaps your work relates to a place of local interest (a major research lab) or national interest (Yellowstone National Park). Here's a section of a story from the *San Diego Union-Tribune* about the Colorado River:

"On April 4, a coalition of government agencies and water and power suppliers plans to hold a ceremony at Hoover Dam in Nevada to unveil a $626 million, 50-year conservation blueprint. One major objective is to improve survival chances for razorbacks and 25 other species in the lower Colorado River."[11]

9. Is it linked to an anniversary? It is quite commonplace for reporters to link stories to anniversaries (especially in multiples of fives). If you start looking for these linkages in the news, you'll notice how often they occur. For example, if you've developed a more efficient electric motor, Edison's 160th birthday, February 11, 2007, would be a perfect time to pitch the story. And here's an excerpt from one of the many stories on the 25th anniversary of Mount St. Helens' eruption:

"The eruption of Mount St. Helens 25 years ago was the must-see event of the century for many people unless the ash cloud meant they couldn't see anything at all.

'I was 13 in 1980, and I was watching it in Massachusetts,' Vancouver geologist Seth Moran said. 'I was riveted to the TV screen.'

Mount St. Helens has shown recently that it can still draw crowds and cameras.

But volcano-watching now includes a study of the unseen. Moran and other U.S. Geological Survey scientists can monitor volcanic activity from a distance, using a toolbox of instruments that has been updated since St. Helens made history on May 18, 1980."[12]

10. Will it touch people's emotions? Ever wonder why a whale stranded on a beach is always a top news story? It's not because it has national implications. Television producers know the folks back home watching the news might feel sad for the animal. We're not suggesting that you manufacture some event that pulls at the heartstrings. But if you have a story to tell that is funny, sad, touching, or even gross, the reporter is more likely to consider it newsworthy. If your subject has a name, it's even better, as evidenced in the first paragraph of this *Boston Globe* article, which is also a little graphic:

> "Scientists may never know what caused the death of a humpback whale named Beacon whose carcass washed ashore on Newcomb Hollow Beach over the weekend because Cape Cod researchers found the mammal's internal organs had deteriorated extensively."[13]

LESS IS MORE

For your initial meetings—and in general—it is much better to err on the side of restraint in claiming that something is newsworthy. After your first meeting, you can still call or email the reporter if you're not sure it is "news." Indeed, an important part of your work with the media is feeding reporters information. That's the currency that sustains your relationship with reporters. Don't say out loud that you've "got big news" if there is any question in your mind. Instead, just make the best pitch you can—highlighting the criteria discussed above—and let the reporter decide. Then, in your talks with reporters, focus on your main messages and submessages.

TAKING IT TO THE NEXT LEVEL

Our guess is that you will find reporter meetings quite illuminating and probably a bit of fun. Consider setting a goal to initiate relationships with a couple reporters every year, but more if you can. As we've suggested, start with a reporter from your local newspaper. Then six months from now (or sooner, if you have time), call and arrange a meeting with an editorial writer at the local newspaper, a reporter with the NPR affiliate in town, or a reporter from the local television station.

Many reporters use local outlets as stepping-stones to larger media outlets. In a few years, the reporter who once wrote for the *Dallas Morning News* might go on to write for *Newsweek*; the reporter who covers your state university might move up to Good Morning America. You never know, so try to make these relationships last. And if a reporter does move on, be sure to get to know his or her replacement.

Once you've developed good relationships with local reporters, you can start to think about getting to know reporters at the regional or national level. (The Resources section of this book lists general contact information for most major media outlets.)

The key here is to prioritize. If you live in Iowa and study prairie grass, it probably makes more sense to reach out to reporters at papers in other Plains states than to give the *Washington Post* a call, unless you have a story that is truly unique or national in scope. Likewise, if you work on an issue with national or global significance and have startling visuals, CNN is probably a more appropriate choice than a television station in Montana. Just keep in mind the limitations on a national reporter's time. NPR's Richard Harris says, "Many science reporters, myself included, are generalists, so it's impossible to develop a lot of relationships with people in every possible field of interest. I'm busy doing NASA one day and stem cells the next." Still, Harris has cultivated scientists as sources and calls them when issues arise in their field. If you do contact a national reporter, be sure that your pitch is relevant and interesting to that reporter's audience, and familiarize yourself with the outlet's coverage to be certain how your issues or views might fit into their stories.

Keeping Up the Lines of Communication Meeting reporters in person or over the phone is a big first step in developing a working relationship. But this is just the beginning. Update the reporter on developments that you think might interest him or her. Give the reporter a "heads up" when an interesting event is taking place or a compelling story is on the horizon. Let the reporter know you're available for comment if a story in the news is related to your expertise. Email her or him when something is happening nationally that the paper should cover or, alternatively, pick up from the Associated Press or Reuters. At minimum, try to contact the reporter four or five times a year, just to stay on the radar screen, even if it's just a very brief email.

WILL REPORTERS EVER BE INTERESTED IN MY ISSUE?

For those of you who might feel that your specialty is just too arcane for the news media, you can still have an impact. By developing a relationship with a local reporter, you can potentially serve as a self-appointed advisor to the newspaper, letting a reporter know when local scientific developments are taking place, or encouraging the paper to run a wire story on a national scientific issue. Newspapers care what their readers have to say. And if their readers clamor more frequently and loudly for science news, they will do a better job of delivering.

CHAPTER 7

Choosing the Right Communication Tools

Now THAT YOU know how to talk about your research or views, you need to figure out the best vehicle for getting your message out to the public. There are many different ways to communicate with the press, ranging from formal (editorial meetings) to informal (email correspondence), complex (video news releases) to simple (letter to the editor). We refer to these communication techniques as "tools," and which tool you choose depends on a variety of factors, including your message, your target audience, and your budget.

In our conversations with scientists, we've learned that too many of you rely on only one mode of communication with reporters. Some only communicate through press releases that their media relations office generates, while others only feel comfortable writing letters to the editor or op-eds. Many more just wait until reporters call them. While that is certainly a start, if you are going to reach the largest audience possible, you need to use all the communication tools at your disposal (or at least more than one or two). In this chapter, we discuss these tools in detail and explain how you can put them to work to effectively communicate your messages.

CULTIVATION MEETINGS

As the last chapter highlighted, the best place to start your media work is by establishing yourself as a source with reporters in your local area. We mention this here again because, as many of the scientists we interviewed affirmed, developing working relationships with local reporters is one of the most effective ways you can have a positive impact on media coverage. You should maintain these relationships by updating reporters regularly through phone calls and emails. Once you've developed a good rapport with your local press, start thinking about what other media outlets in your region might be interested in your work. If you work on an issue that is national in scope, peruse the websites of larger media outlets to see which reporters might be interested in your research or views, keeping in mind that you have to meet a higher threshold of newsworthiness to attract their interest.

EDITORIAL BOARD MEETINGS

Once you've met with a local reporter, set up a meeting with an editorial writer at your local newspaper. These anonymous writers' opinion pieces represent the views of the paper's editorial board, a group of writers and senior managers at the newspaper. Editorials appear next to letters to the editor on the left-hand editorial page.

Bruce Dold, editorial page editor of the *Chicago Tribune*, says that while his paper's reporters cover news objectively, it is the job of editorial writers to "synthesize and interpret those events, and often to suggest what we see as logical solutions for the dilemmas that vex our times."[1] As scientists with expertise on many of today's most pressing dilemmas and solutions, you can help sway an editorial writer's opinion.

Editorials can be very influential. They are read not only by the general public, but also by local, state, and federal legislators, as well as business leaders. If you are releasing new research, a favorable editorial from a newspaper can raise the profile of your work by telling people why they should pay attention. If you want to raise the visibility of a policy issue, an editorial can help persuade the public why it should care.

To arrange a meeting with an editorial writer, simply call the main switchboard of the newspaper and request to be transferred to someone in the editorial department. Once you're connected, ask which editorial writer is most likely to write about science-related issues. Ask for that writer's contact information and send him or her an email that briefly summarizes your research or policy issue and why it's important. Request a meeting and offer a few possible meeting times. The editorial writer will most likely reply and work with you to figure out a time to meet. If you don't get a response, follow up with a phone call.

It's possible the editorial writer will suggest you meet with the full editorial board. If you are given that option, jump on it—the more people on the board you can educate, the better. Editorial meetings are different from reporter meetings in that your goal is to make clear what you would like the paper to weigh in on and then make the case. Just as you would do for a regular news article, think ahead about how you would like the headline of an editorial to read. An editorial headline should have an opinion, highlight a problem, or propose a solution—and it should be the main message you deliver in your meeting. Your other messages make the case for the headline and include a call to action—whether it is to pass a law, undertake (or stop) a project, request funding, urge more research, or simply raise public awareness. Take some background information to the meeting to give to the editorial writer or editorial board members. This information (possibly as part of a press kit, which we discuss later in the chapter) should provide the details of your issue so you can focus on your talking points in the meeting.

Editorial meetings typically last an hour or less. After you've made introductions, make clear why you'd like the newspaper to weigh in on your issue, leaving plenty of time for questions. When the meeting is winding down, you may have a sense of whether the paper is interested in writing an editorial. However, don't ask point blank if someone will write about the issue. If the editorial writer/editorial board doesn't seem interested, don't worry. Like a reporter cultivation meeting, you should view this as just the beginning of a long-term working relationship. You can contact the editorial writer in the weeks or months ahead to follow up on this or other issues. And remember, editorial writers are looking for ideas for the future just like reporters, so be judiciously persistent (i.e., don't call every week, but don't call only once a year either).

PRESS RELEASES

Visit any newsroom in America and you would be astounded by the number of press releases reporters receive. Tom Avril, an environmental reporter at the *Philadelphia Inquirer*, says his paper's science and health desk is "inundated with news releases every day." These days, it seems press releases are issued for just about anything, from promotions to conference announcements to the awarding of a small grant. Gareth Cook of the *Boston Globe* observes, "My sense is that a lot of press releases are sent out for bureaucratic reasons—like a dean who is really excited about a new building opening—rather than news reasons." Cook adds that he pays "almost no attention to these press releases, because they generally do not contain anything that is newsworthy."

Roger Johnson, a biochemist who later became a science writer, founded Newswise, an online news service that distributes press releases to thousands of journalists every day. Having read more than 30,000 press releases, he knows that an effective press release contains information that is compelling to reporters. "Scientists need to think carefully about the content," Johnson says. "They need to ask, 'Is it a story that the media are going to be interested in?' A lot of news releases come from the point of view, 'I want that information out there' or 'this is important, and people should be interested.' That is not considering the reader. It comes from a patronizing point of view. If you really want to connect with people, you have to consider your reader and what will interest them. The readers of news releases are reporters. They are interested in news."

Gareth Cook reiterated this point. "The main piece of advice I would give," he says, "is that [scientists] spend a little bit of time thinking about whether [their press release] is really, truly newsworthy. Can they imagine reading about it in the publication that they are planning on sending it to?" Cook says press releases on science issues need to "state clearly, without jargon, what is really new, and why it matters. Why should a reader care about this?"

As scientists, you may be under pressure to issue press releases from your university's media relations department. Many well-intentioned public information officers (PIOs) try to gain media attention for their school or organization by sending out frequent press releases, even if a topic isn't necessarily newsworthy. This approach, however, could eventually backfire. It's like the story of the boy who cried wolf. If a PIO cries "News! News!" to reporters and follows up with press releases that aren't newsworthy, reporters will soon ignore the PIO, and when the PIO actually does have news, the reporter may not pay attention.

Our goal is to make sure you use press releases as effectively as you can because a well-conceived and well-written press release *can* garner media attention, and potentially a lot of it. When you have news, the press release is one of the best tools to communicate it to the public.

Press releases should alert reporters about a possible story to cover. While many reporters view press releases as teasers and will contact you for more information, some reporters will create stories directly from a press release. You can write a press release about new research, a surprising development that touches people's emotions, a policy development, or any of the other combinations of characteristics that make a subject or an event newsworthy. If you are uncertain whether your research is newsworthy or not, talk to friends or family members about your work. Their questions might draw your attention to a new angle of your story or something you overlooked. The bottom line is that press releases should provide reporters with the key elements of an interesting or compelling story so they can write an article or produce a broadcast segment based on the information presented.

Elements of an Effective Press Release An effective press release combines the basic components of a story in the media (i.e., what is new, why it is important, where and when it is happening, and who is involved) with sound bites from someone who has expertise related to the subject. The release should essentially be your main messages and talking points (submessages) in the form of a news article. In addition to this key information, there are several other important elements:

The "Hook." "The key to a good press release is the lead," says Roger Johnson. "It always comes down to, 'What is the essence of what's new in this story?'" Start with an attention-grabbing headline that you would want to see in the morning paper, then an introductory paragraph of two to three very short sentences that highlights the news you are trying to convey. Says Johnson, "I encourage scientists to identify the core—what is important, unique, new and interesting. Get that essence in one or two sentences. Make it understandable but also interesting."

Take a look at the press release on page 117 from the Wildlife Conservation Society (WCS) at the Bronx Zoo. It was Stephen C. Sautner's job to make a scientific paper published in the *Journal of the Zoological Society of London* understandable and resonant to the media. Sautner, the assistant director of conservation communications at the WCS, says the report was "your typical dense scientific study, written exclusively for a scientific audience." The paper was titled *Rapid ecological and behavioural changes in carnivores: the responses of black bears (Ursus americanus) to altered food.* "A title," Sautner says, "that does not exactly roll off the tongue. Our job was to translate this hard science into something readable."

Sautner read through the report trying to find information that people could relate to. One line caught his attention. "Individuals at urban interface areas relative to wildland conspecifics were: (1) active for significantly fewer [hours] per day (8.5 vs. 13.3 h; P < 0.01)..." While this would be over the head of the general public, Sautner had an idea. "After consulting with the authors," he says, "we learned that this meant very simply that urban black bears ate their fill in fewer hours, mostly by raiding dumpsters, then remaining relatively inactive." And just like that, the headline "Urban Black Bears Becoming Couch Potatoes, Study Says" was born. Sautner says it was a "simple, humorous way to communicate" one of the study's main findings in a way that would resonate with the public.

Journalistic Style. Press releases should be written in a style similar to a news story, where the most important facts come first. In journalism parlance, this style is called the inverted pyramid—the big picture grabs a reader's attention and sets the context, then each subsequent paragraph gradually narrows to smaller details. In other words, focus on the news and not on the university releasing it. Edie Lau of the *Sacramento Bee* recommends that "the names of people and institutions who the press staff thinks need to be credited . . . can come at the end" of the press release. Remember to write in the active voice and, like the WCS press release, aim for short paragraphs with easy-to-understand language.

• Press release begins with a headline and first paragraph that concisely encapsulates the story.

• Body of press release includes supporting talking points and main sound bites.

• End of press release includes background details and final quote.

FIG. 2: THE INVERTED PYRAMID

Sound Bites. In chapter 4, we describe the different types of quotable sound bites scientists can use. Pepper similar sound bites throughout your press release. Although the first sound bite usually comes in the second paragraph, if the second paragraph is short as it is in the WCS press release, then the sound bites can be placed in the third paragraph. A "set" of sound bites comprises one short pithy sentence in quotes, followed by the name of the expert and her or his organizational affiliation, then one additional short sentence. Here's an example from the WCS press release.

"Black bears in urban areas are putting on weight and doing less strenuous activities," said WCS biologist Dr. Jon Beckmann, the lead author of the study. "They're hitting the local dumpster for dinner, then calling it a day."

Try to include another set of sound bites after the fourth paragraph, and if it makes sense, at the end. If a reporter calls you up to discuss the press release, weave in the quotes from the press release as you answer questions. (Don't read them, just speak naturally.) If your sound bites aren't pithy, the reporter will try to entice quotes from you, and if you start thinking off the cuff, you may well end up seeing a quote you don't like in the media.

Brevity. Keep press releases to one page. It can be challenging to summarize a large study in one page, but it can and should be done. It will, in turn, steer the reporter to the points you want to make. A three-page press release would have so much information in it that the reporter could take the information and go in several different directions—some of which might not be what you intended—or, worse, he might not read it at all. Reporters are pressed for time. They want to look quickly at your press release and decide if it is of interest to them. If it is, they can always call for more information and perhaps read your report.

Date and Contact Information. If a reporter received an email from you today containing a press release, he or she would expect that the news in the press release could be reported on immediately. To make it clear that is the case, a press release should contain FOR IMMEDIATE RELEASE on the upper left-hand corner of the page with a date listed both underneath it and in the dateline. (The dateline lists the city where the news is coming from and the date, and is listed before the first paragraph.) That way, reporters know they can cover the story immediately, and if they stumble across it later during a web search, they will know when it was released.

Sometimes PIOs or journals will "embargo" press releases and reports, meaning that a reporter can't publish any of the information contained in the release until a specified date and time. Embargoed press releases are used by *Science*, *Nature*, and other journals to release information equitably to reporters across the country. Other institutions will occasionally embargo information related to a large study that would be difficult to read and report on in the same day; the embargo gives reporters extra time to prepare a story.

Be sure to include in the press release the names and numbers of one or two people the reporter can contact. Often these people will be media relations staff at a university or institution. However, reporters most appreciate when the phone numbers and email addresses of scientists are included on the press release. "That's more efficient than our having to call the press staff to obtain the contact information," says Lau. On the other hand, if you are new to media work, you may prefer not to have your name listed as a contact. That way your PIO can serve as a facilitator and find out everything you'll need to know before the interview, which will enable you to be more prepared.

Always include a website somewhere on the press release so reporters know where to find more information if they're interested. If the website is not a direct link (i.e., reporters will have to type in the web address), try to keep the URL short. It may make more sense to direct them to a home page with a link to your report rather than a long report URL.

Consistent Format. Printed copies of press releases (i.e., press releases you will distribute by hand or by mail) should include a logo or letterhead at the top and some indication that it is a news release. Some institutions prefer to double space their press releases while others single space with a double space between paragraphs. Some write the headline in all caps while others use a large font. Some include a subhead line while others don't. Some signal the end of a press release by using three pound signs (# # #) while others use the number thirty offset by dashes (-30-). Either way, the notation should be center justified.

In the end, it doesn't matter which formatting style you use, as long as you use it consistently. What matters most is that you have something newsworthy to share with reporters, you summarize your main messages with an effective headline and first paragraph, and you have quotable sound bites that relate to your key messages and talking points (submessages).

Ample Lead Time. A press release advertises your research or views to the media, and thus, the world. It doesn't make sense to spend a year writing a report but only an hour writing a press release. It can take some time to decide what your main messages are and how you should express them. Writing a press release on a collaborative study can take even more time, as your colleagues might have different opinions about how to emphasize certain aspects of the research, might suggest alternative sound bites, or might even disagree on the overall message and headline. For these reasons, you, your media staff, and any colleagues involved should discuss the main messages and talking points well before the press release is drafted. It will make the press release writing process much smoother and ensure that the proper story gets told.

The black bears press release from the WCS, for example, was a collaborative effort between Stephen Sautner and Dr. Jon Beckmann. "I remember specifically telling Jon to 'trust us' on this headline, and he did," says Sautner. "I should point out that virtually every article I can think of that ran did an excellent job in reporting the scientific underpinnings of this important conservation story."

NEWS RELEASE

CONTACT:

Stephen Sautner (718-220-3682; ssautner@wcs.org)
John Delaney (718-220-3275; jdelaney@wcs.org)

*URBAN BLACK BEARS BECOMING
COUCH POTATOES, STUDY SAYS*

NEW YORK (Nov. 24)—Black bears living in and around urban areas are up to a third less active and weigh up to thirty percent more than bears living in more wild areas, according to a recent study by scientists from the Bronx Zoo-based Wildlife Conservation Society (WCS).

The study, published in the latest issue of the *Journal of Zoology* says that black bears are spending less time hunting for natural food, which can consist of everything from berries up to adult deer. Instead, they are choosing to forage in dumpsters behind fast-food restaurants, shopping centers, and suburban homes, often eating their fill in far less time than it would take to forage or hunt prey.

"Black bears in urban areas are putting on weight and doing less strenuous activities," said WCS biologist Dr. Jon Beckmann, the lead author of the study. "They're hitting the local dumpster for dinner, then calling it a day."

In addition, the authors say that urban black bears are becoming more nocturnal due to increased human activities, which bears tend to avoid. Bears are also spending less time denning than those populations living in wild areas, which the authors say is linked to garbage as a readily available food source.

The authors suggest that as humans continue to expand into wild areas, and as bears colonize urbanized regions, people must be educated to reduce potential conflicts. Local ordinances should be passed mandating bear-proof garbage containers for homes and businesses.

"Black bears and people can live side-by-side, as long as bears don't become dependent on hand-outs and garbage for food," Beckmann said. "Lawmakers should take a proactive stance to ensure that these important wild animals remain part of the landscape."

###

Copies of the study are available through WCS's Conservation Communications Office (1-718-220-3682).

Disseminating Your Press Release The best way to get your press release to a reporter is to email it to him or her directly. If you work at a university or large institution, your media relations office should be able to assemble a media list with reporters' names and email addresses and send out the press release. If not, you should maintain your own list of media contacts and send the release yourself. Hopefully, though, you've already been thinking about which media outlets—both local and national— might be interested in your work.

If you have compelling images or graphics related to your story, mention it in your email. The same goes for any sounds or smells (anything visceral that either translates directly to the media or can be described). And be sure to embed the text of the press release in the email. "Don't send the [press release] as an attachment," advises Edie Lau. "I don't have the time or patience to click and click to find the gist of the pitch."

If you don't have access to a PIO or media relations staff, there are a number of paid services that can help you distribute your press releases. Newswise sends research-related science news to thousands of science reporters around the world. EurekAlert!, run by the American Association for the Advancement of Science, also sends press releases on science-related subjects to science reporters around the globe. If you have a more general story to tell, U.S. Newswire and PR Newswire can disseminate your press releases to assignment editors and radio and television producers, as well as reporters around the country. (Contact information for press release distribution services is located in the Resources section of this book.)

Sautner of the WCS sent the black bears press release to more than 100 reporters from his own media list, but he also posted the release on Newswise and EurekAlert! "The media coverage of this story was astounding, and to date represents one of the most-covered stories WCS has generated," Sautner says. Indeed, the story generated print and broadcast coverage around the world.

In the subject line of the email to reporters, use an abbreviated headline that captures the news in five words or fewer. Many reporters ignore or delete emails with subject lines that aren't compelling. For example, the subject line for the WCS could read: Black Bears Becoming Couch Potatoes. The press release text should have no formatting other than left-justified text and one line space between paragraphs. The top section should include the release date, organization name, and contact information.

Try to email your press release as early in the day as possible. The earlier you send it out, the more time the reporter has to prepare a story. If you can, avoid sending press releases on Monday and Friday (never send them over the weekend unless the news relates to an emergency). The first day of the week is often extra busy for reporters while on Friday many are trying to wrap up their week or preparing stories for the Sunday paper.

PRESS STATEMENT

While the press release is the best tool to use when you are releasing your own news story, the press statement is a more efficient way to send your talking points to the media when you are commenting on someone else's research or a policy story already making news. A press statement is what the name implies—a short statement from you about a newsworthy topic. Like a press release, the press statement includes a headline, contact information, and dateline. Unlike a press release, which is written in the style of a newspaper article, a press statement contains your main messages and sound bites written in a few paragraphs. The press statement should be used when the press is covering a newsworthy event that reporters already know details about. You are simply providing a series of comments and quotes from which the reporter can pick and choose. If she or he is unaware of the subject of your statement, the reporter would be better served by a press release which provides background information.

Here is an example of a statement issued by the Union of Concerned Scientists (UCS) for Dr. Kurt Gottfried, emeritus professor of physics at Cornell University and chairman of the UCS Board. Note that the statement (in email format) is about another study that had recently been released.

FOR IMMEDIATE RELEASE:
November 17, 2004

UNION OF CONCERNED SCIENTISTS
CONTACT: Suzanne Shaw, (617) 547-5552

NATIONAL ACADEMY OF SCIENCES FINDS
POLITICAL QUESTIONS INAPPROPRIATE
Scientists Should Not Be Asked Party Affiliation, Voting Record

STATEMENT BY KURT GOTTFRIED, CHAIRMAN,
UNION OF CONCERNED SCIENTISTS

In a report released today on the presidential appointment process, the National Academy of Sciences strongly stated that it is inappropriate to ask scientists and other technical experts their political party affiliation, voting record, or personal positions on particular policies when considering them to serve on federal science advisory panels.

The report echoes the concerns voiced this year by more than 6,000 scientists–including many Nobel laureates, National Medal of Science recipients, university presidents, and leading medical researchers–that nominees for science advisory panels should be judged only on their expertise and professional qualifications. To do otherwise undermines the integrity of scientific input into policy decisions and can compromise public health, safety, and the environment.

The overwhelming majority of Americans agree with the scientists. A recent national survey found that two-thirds of the public felt it is not acceptable to ask about party affiliation or recent presidential voting when considering a candidate for an advisory committee.

Congress should move forward to strengthen and enforce rules governing appointments to scientific advisory panels to forbid this improper line of questioning. Now that the constraints of the campaign season are over, the administration should follow the NAS recommendations and move quickly to restore the stature of the director of the Office of Science and Technology Policy to assistant to the president.

###

CALLS

Usually there are a few media outlets that are your leading choices for carrying the story. Perhaps the reporters are local and you've already established a relationship with them, or maybe the reporters work at an influential national media outlet like the *New York Times* or *Wall Street Journal*. In either case, if you want to ensure reporters read your press release, you should call them after the press release has been sent. There is a chance they will read it without any additional input from you, but given the volume of emails reporters receive, you should call to alert their attention to it. For example, if you were Dr. Beckmann at the WCS, you could call with this short message: "Hello, this is Jon Beckmann. I'm a scientist at the Bronx Zoo-based Wildlife Conservation Society. We sent you a press release this morning on a new paper we just published in the *Journal of Zoology* that shows black bears living in urban areas are becoming couch potatoes. The bears are up to a third less active and weigh up to thirty percent more than bears living in wilder areas. If you didn't see the release, or if you have any questions, please don't hesitate to call me at..."

When you have a breaking news story—something that is timely and meets the newsworthy threshold of being unique, relevant, and interesting—you should never hesitate to call reporters. "I don't get a lot of scientists calling me with hot tips," says NPR's

Richard Harris, "and actually it would be nice to get more of them." As with press releases, it's best to call newspaper and television reporters early in the day (definitely before 3 p.m.), unless you have major breaking news to share. In most cases you can call radio reporters any time during the day, but be aware that they could be working on a story for the evening or morning news.

EMAILS

From time to time, you may frequently come across information that doesn't merit a press release but might be useful to a reporter. In situations such as these, we encourage you to email reporters short messages giving them a "heads up." Many of these "heads up" emails might well lead to news stories. Edie Lau says she pays "particular attention to personal pitches from reputable scientists." These are not news releases, she says, "but notes written directly to me with their thoughts on why they think a particular subject is newsworthy."

In fact, as we discussed in the last chapter, one of the best ways to become a reliable source to reporters is to email them short, regular updates on either your research or policy issues you are tracking. Reporters we've talked to stress that these messages should be brief and to the point. Gareth Cook says, "I much prefer short, cogent emails that tell me what is new, why it matters, and how to reach the principals if I want to." Also consider sending a reporter a one-sentence note after a particularly good story, or if you think an article was incomplete or inaccurate (never insult the reporter when informing him or her of an error). Your goal is to make sure that the reporter has all the facts. Be objective, reasonable, and brief.

LETTERS TO THE EDITOR

The letters to the editor section is one of the most popular sections of a newspaper. As a result, writing a letter to the editor (LTE) can be an effective way to communicate your talking points and views directly to the public. Every newspaper has a letters section, usually located on the editorial page. Editorial page editors like short LTEs because they can fit many of them on the page. Newspapers such as the *Los Angeles Times* and *Washington Post* print anywhere from six to eight LTEs a day. The *New York Times* also includes LTEs in its Tuesday science section ("Science Times"). The odds of having your LTE published vary widely depending on the size of the newspaper. Art Coulson, editorial page editor of the *St. Paul Pioneer Press*, says his paper has "room to publish about one out of every four letters we receive."[2] The *Chicago Tribune*, on the other hand, receives thousands of LTEs but can only publish about seventy per week. "Competition is fierce," says Dodie Hofstetter, who handles LTEs for the *Tribune*.[3]

There are four reasons you might want to consider writing an LTE:

1. *Amplify a message.* If you read a well-written or well-researched news story, editorial, or op-ed on an issue you care about, your LTE can add another angle or explain the implications or stakes of the story.

2. *Rebut views.* If a writer presented a balanced news story, but your opinions differ from the person being quoted in the article, your LTE can present an alternative opinion.

3. *Point out an important mistake.* If you encounter an error in a news story, your LTE can provide corrected information. (If you were a source for the story and the error is large, ask the reporter to issue a correction. If you were a source and the error is not large, write a short note directly to the journalist pointing out the mistake.)

4. *Make a statement about lack of coverage.* If your local newspaper is ignoring a major science story, your LTE can gently chide the newspaper for ignoring the issue. Even if the LTE isn't published, you can help influence the paper's subsequent coverage of science issues.

The key to effective LTEs is to have an opinion. Peter J. Wasson, opinion page editor of the *Daily Herald* in Wausau, Wisconsin, says that people in his town want to "read the views of their neighbors." The letters he prints are "original, thought-provoking submissions on topics of interest."[4]

Strive to keep your LTE under 150 words, and never go over 250 words, even if your paper prints letters this long (the longer they are, the more likely they will be edited, potentially cutting out your most important messages). Your newspaper should list an email address for submitting LTEs. If you can't find it, check the paper's website or call the main number and ask for the information.

Here is an example of an effective LTE published in the *New York Times* on March 24, 2005. The fact that the letter is terse makes it all the more powerful.

TO THE EDITOR:

Re "A New Test for Imax: The Bible vs. the Volcano" (news article, March 19):

It is a sad day when marketing studies and pressure from religious fundamentalists censor science from science museums.

Evolution in a billions-of-years-old universe that began in the Big Bang is the central framework of modern science. A science museum that does not have the courage to show a film that discusses that central framework is not worthy of the name.

Jon Machta
Amherst, Mass., March 19, 2005

The writer is a professor of physics at the University of Massachusetts at Amherst.

Notice that the writer is identified as a scientist. When you submit your LTE, be sure to include information about your scientific, academic, or professional affiliation and expertise. You may even want to allude to your credentials in the text of the letter.

One word of caution that may seem obvious: never submit an LTE under your own name that someone else wrote. Always use your own words, even if a friend or an organization has made suggestions on what you may want to say. "Letters editor Amira Awad and I routinely cruise the campaign and advocacy group websites and make note of the form letters they contain," says Art Coulson. "The National Conference of Editorial Writers posts copies of form letters its members have received—several different letters each day. And it's pretty easy to smell a rat when we receive dozens of identical letters signed by different 'writers.'" Coulson says, "We just ask that you express your opinion in your own words. And please include your REAL name, address and phone number. We do call to verify a letter's origin—if we can't reach you because you've given us bum information, your letter is heading for the shredder."[5]

OP-EDS

Op-eds offer you a chance to communicate your viewpoint in a visible way directly to the public, with many more words than an LTE. They appear opposite the editorial page in most newspapers (hence the "op" in "op-ed") and are often located next to columns by syndicated writers (such as George Will or Molly Ivins). Op-eds are articles written by citizens not affiliated with the newspaper—anyone from business executives to scientists to politicians to parents and even school kids.

The most important step in writing an op-ed is deciding whether you should write one in the first place. William A. Collins, who heads the op-ed distribution service Minuteman Media, suggests that a "useful preparation for op-ed writing is to try to explain the subject matter to your barber, manicurist, or brother-in-law. This will show you whether their eyes glaze over, and you may discover certain phrasing or analogies that strike a chord. Or it may cause you to reassess your whole plan."

How to Write an Op-Ed Just like a good essay, op-eds have a title, introduction, body, and conclusion.

The title is the first thing a reader will see, so you want it to catch the reader's attention and entice him or her to read what you have to say. It should also reflect the theme of your piece so the reader can get a sense of what you're writing about. Keep in mind, though, that editors might change your title for layout reasons or to make it even pithier.

The introduction, like the title, should pique a reader's interest and encourage him or her to keep reading. This is where you first convey your overall message or position, using timely references, colorful language, or metaphors to get the reader's attention. Try to limit the introductory paragraph to three short sentences.

The body further develops the main point you are trying to make. Paragraphs should be short and focused, ranging from three to five sentences. Try to make one supporting point in each paragraph, and be sure each paragraph flows smoothly into the next. Every paragraph should tie back to the introduction and your overall message. Avoid going off on tangents.

The conclusion wraps up the piece, tying together all the threads of the previous paragraphs. You should restate your overall point or opinion in the conclusion; your last sentence should be memorable to help make your message stick in the reader's mind.

To make your op-ed effective and increase its chances of being published, keep the following details in mind.

Length. Editorial space is often quite limited, so it's important to check the length requirements of the newspaper you are targeting before you start writing. Ideally, an op-ed should be 500-700 words, though some editors, such as the *Providence Journal* editorial page editor Robert Whitcomb, will consider pieces up to 1,000 words.

Contact information. Don't forget to list your contact information at the end of your op-ed, including your name, university or organization, title, work and home phone numbers, and address. Many papers will not run a piece without first being able to verify the author.

Focus. Gloria Millner, op-ed editor for the *Cleveland Plain Dealer*, finds that scientists often "write op-eds that are too long for use on a daily op-ed page. . . . They try to write about too many concepts, instead of focusing on one aspect of an issue." For example, she says, "a scientist may want to correct some widespread misunderstanding about stem cell research. When the subject is hot (i.e., there are a lot of stories on page one, or it's an election debate topic, or something like the death of Christopher Reeve has brought the issue back to the surface), [an editor] can probably be persuaded to use a longer piece. But if the issue is not quite so hot, it's best to focus on something that can be made understandable in 700 words. For example, there may be a dozen common misunderstandings about stem cells. Why not focus on one that seems most prevalent or most dangerous?"

Language. Most of your readers are not experts in your area of research, so it's important to write for a general audience. Bob Ewegen, deputy editorial page editor at the *Denver Post*, says his "key recommendation would be the KISS rule: keep it simple, stupid. Don't throw around a lot of alphabet soup or undefined jargon."

Local Impact. Millner advises scientists to look "at the industries or research facilities in the paper's town. Here in Cleveland," she says, "we're very interested in the work being done at local universities—for example polymers, liquid crystals."

Ed Williams, editorial page editor of the *Charlotte Observer*, echoes this point, saying "failure to take opportunities to localize or regionalize issues" is one of the most common mistakes of scientists who write op-eds. For his readers, Williams wants to know, "What does [the issue] mean to the Carolinas? To Charlotte?"

Opinion. Last, but certainly not least, be sure to have an opinion—whether it's on the value of the research or the extent of a problem or the steps that need to be taken to solve the situation. "Take a clear stand on the public policy choices," says Williams. "The scientific issues may be fraught with subtleties, but the political choice usually is either/or. The writer could also explain what the public policy choice should be, as opposed to the way it has taken shape in the political process." Williams says he frequently turns down op-ed submissions because they fail "to express clear (to op-ed page readers) views on timely topics and important (or interesting or both) issues."

The following op-ed, "Missile Defense in a Vacuum," was written by Lisbeth Gronlund and Kurt Gottfried, scientists from the Massachusetts Institute of Technology and Cornell University, respectively. (Gottfried is chairman of the Union of Concerned Scientists; Gronlund is co-director of UCS's global security program.) The piece was published in the *Washington Post* on May 3, 2000, and is a good example of a no-nonsense op-ed about a scientific matter related to a policy question. The scientists wrote about a complicated subject with language that is easily understandable, yet not dumbed down.

In the first paragraph, the writers take an immediate position: that the missile defense system won't work. They then describe what's at stake and summarize their research, which supports the central theme introduced at the beginning. As they make their case, the scientists allude to opposing views as they rebut them. And, if you are a scientist who is loath to show any emotion in your writing, pay particular attention to how Gronlund and Gottfried communicate strong opinions in a staid manner (though they do manage to include language that paints a picture, e.g., "...the Pentagon has no way of knowing whether its

toddler will be able to survive its first run through a mine field.")
Finally, the piece ends with a proposed solution: "The president
should only move forward with deployment of a system that
has been successfully and repeatedly tested against a credible
threat." The op-ed was not only well written, but, equally important, it was timely.

MISSILE DEFENSE IN A VACUUM

President Clinton's impending decision on whether to begin
building a national missile defense system has produced a
heated debate: How much will this system cost? What effect
will deployment have on U.S.-Russian nuclear reductions?
Few are asking the basic question: Will this system work? The
answer is no.

The mission of national missile defense is to protect the
United States from attacks by emerging missile states such
as North Korea, which could in the future acquire long-range
missiles armed with nuclear, biological or chemical weapons.

The United States is the world's strongest military power
with the most advanced technology base, whereas North
Korea is one of the world's poorest countries. Surely, you will
say, missiles launched by such countries could not penetrate
sophisticated American defenses.

They could. For even if the planned U.S. missile defense
technology worked perfectly (a big if), the system could be
defeated by relatively simple steps, called countermeasures,
that would foil or overwhelm the defense. And such countermeasures are easier and cheaper to develop and deploy than
long-range missiles.

This conclusion is based on a recent study we co-
authored with nine other physicists and engineers, half
of whom have served as senior consultants to the Defense
Department. We described three specific and readily avail-

able countermeasures in detail. Then we asked whether an attacker could defeat not just the first phase of the defense that the United States would deploy by 2005 but also the complete system slated for 2015, with all of its planned radars, satellite-based sensors and ground-based interceptors. Using publicly available information about national missile defense, and assuming the defense is constrained only by the laws of physics, we found that an attacker using any one of the countermeasures we examined would defeat the fully deployed missile-defense system.

For example, the attacker could hide its nuclear warheads inside aluminum-coated Mylar balloons and release them along with dozens of empty balloons. Because none of the missile-defense sensors could tell which balloons contained warheads, the defense would have to shoot at all the balloons. Even a small attack could have more than enough balloons to exhaust the supply of interceptors.

An attacker using biological weapons could also defeat the defense. To be effective, biological agents have to be distributed over large areas. We showed that an attacker could readily do so by separating each missile payload into a hundred or more small "bomblets." As a bonus to the attacker, such bomblets would overwhelm the defense.

These problems are systemic because the planned missile defense relies on interceptors that must hit targets directly in the vacuum of outer space. This design cannot be modified to make it effective against such countermeasures.

Some supporters of the missile-defense system argue that countermeasures are too difficult for a country like North Korea. They ignore the 1999 National Intelligence Estimate by the U.S. intelligence community, which warned that emerging missile states would be able to use "readily available technology" to develop countermeasures and could do so by the time they deployed any missiles.

Others say that while the first phase of the system will not be able to cope with countermeasures, the full system will—that the defense is "learning to walk before it learns to run." But the tests that will be done prior to the president's decision will not include realistic countermeasures, and no such tests are planned until at least 2005. So the Pentagon has no way of knowing whether its toddler will be able to survive its first run through a mine field.

The president should not make a deployment decision based on unsupported assertions by the Pentagon that the national missile defense system can deal with countermeasures. If the Pentagon disagrees with our analysis, it should explain what is wrong with it. And then the Pentagon should demonstrate, in intercept tests, that the system could defeat realistic countermeasures. The president should only move forward with deployment of a system that has been successfully and repeatedly tested against a credible threat.

Lisbeth Gronlund is a physicist at the Union of Concerned Scientists (UCS) and MIT Security Studies Program. Kurt Gottfried is professor of physics at Cornell University and chair of the board of UCS.

The next op-ed, published in the *Chicago Tribune* on May 4, 2004, takes a very different approach. Sandra Steingraber, a biologist at Ithaca College, uses a narrative, conversational style to describe a small town in Illinois grappling with a legacy of pollution problems from a plastics plant. Steingraber doesn't introduce her central point (that she is against rebuilding the plant and advocates an end to the use of polyvinyl chloride) until she's more than a third of the way into the op-ed. This bucks the guidance of op-ed editors, but the piece is still effective because the introduction draws the reader into the issue by painting the image of a small town grappling with an unwanted legacy of toxic pollution, all from a personal point of view (even though the writer didn't live or grow up there). Ultimately, the op-ed succeeds because it is well written, easy to understand, and interesting, and communicates a strong position along with a call for action. It is hard to imagine it being as effective if it had been written using a more traditional style.

ILLIOPOLIS: CENTER OF DISASTER
AN ENVIRONMENTAL HAZARD

The town in central Illinois where I grew up didn't have a lot of newcomers. So, when a family moved in next door, my sister and I watched the arrival of their moving van with mounting excitement. A boy and a girl, roughly our own ages, walked over to say hello.

"We're from Indianapolis," said the girl. She made it sound like a challenge.

Her older brother chimed in. "It's in the center of Indiana." Then he made a joke about there being no Illiopolis, and they both laughed.

Au contraire, I could have said, had I been a more cosmopolitan child. I could have casually pointed south, where, straight down Route 121, lay the real-life city of Illiopolis.

Well, village of Illiopolis, anyway. With a population of little more than 900 souls, Illiopolis may not have the same cachet as Indianapolis, but it does claim to sit at the center of the state. Its founding fathers planned it that way.

On the night of April 23, Illiopolis became central in another way—in the ground zero sense of the word—when the Formosa Plastics Corp. plant blew up, killing four workers, sending into the air a 200-foot "cloud of flame," cutting power, closing highways, turning the entire populace into evacuees, and making headlines as far away as Taipei, Taiwan, and Liverpool, England.

Now that the towering fireballs have been reduced to smoldering piles of plastic goo, talk about what should happen next has begun. Plant managers have vowed to rebuild, and Republican lawmakers have pledged to find state money to help this facility rise from its own toxic ashes. For three big reasons, this would be a mistake.

First, as a manufacturer of polyvinyl chloride (PVC) resin, the Illiopolis Formosa plant is one of Illinois' largest polluters. Two years ago, it was the subject of an article by writer Becky Bradway that documents its long legacy as an environmental saboteur. Her 2002 essay in E: The Environmental Magazine, titled "Ill Winds: The Chemical Plant Next Door," describes the mysterious fish kills, the recurring chemical accidents, the cancers among area residents.

In 2001 alone, the plant released into the air more than 41,000 pounds of vinyl chloride, according to the Environmental Protection Agency's Toxics Release Inventory Web site. Vinyl chloride is a known carcinogen that is also linked to birth defects. In that same year, the Illiopolis Formosa plant was ranked fourth in the nation by the EPA for vinyl chloride emissions.

Second, Formosa Plastics Corp. has a long history of egregious environmental problems. According to the EPA, the groundwater under Formosa's Delaware PVC plant has been "heavily contaminated" with solvents and degreasing agents.

During the 1990s, some commercial shrimpers resorted to hunger strikes and boat-sinkings to bring attention to Formosa's Texas plant for its devastation to the Gulf Coast ecosystem. And its Louisiana plant continues to release frightening amounts of dioxin, according to the EPA.

In Taiwan, the company's environmental offenses have set off demonstrations.

Third, there is simply no future in PVC. Due to a host of health and environmental problems, PVC is being phased out of all kinds of uses, from medical supplies to building materials. Not only is PVC dangerous to manufacture (consider that the deceased Illiopolis workers wore special nail-less shoes so as not to set off sparks when they walked), but it also must be stabilized with heavy metals and slathered with smelly, toxic plasticizers to make it useable.

Vinyl flooring and wall covering have been linked to asthma in children and respiratory distress in office workers. Vinyl blinds add lead to house dust. During house fires, vinyl makes dense black smoke and hydrochloric acid, which can be deadly to firefighters. It cannot be easily recycled.

For all these reasons, PVC has been restricted for use as a building material in some European nations and eschewed by an increasing number of architects and designers in North America. The U.S. Green Building Council is currently considering a credit for "vinyl avoidance" in construction, a move supported by the Healthy Building Network and already accomplished by their Australian counterparts.

Long-suffering Illiopolis deserves to be at the center of environmental health, not environmental disaster. With green building as the fastest growth sector in building and construction and Chicago Mayor Richard Daley's pledge to turn Chicago into a model for green building, how about turning the smoldering ruins of the Illiopolis plant into a manufacturing center of worker-friendly, environmentally sustainable materials?

Sandra Steingraber is a biologist and author of the book "Living Downstream: An Ecologist Looks at Cancer and the Environment." A native of Pekin, Ill., she is on the faculty of Ithaca College.

Steingraber's op-ed illustrates that there is room for creative license in writing op-eds, but this is the exception, not the rule. Most op-eds from scientists should be written according the guidelines we set out earlier in this section.

The following excerpts from a *San Jose Mercury News* writing guide provide additional tips and examples on what to say (and what not to say) in your op-ed:[6]

WRITING TIPS

Support your arguments with a mix of facts, quotes from authorities, and examples.

Example: Since the teen curfew law went into effect a year ago, police report serious crime by juveniles is down by 10 percent; graffiti is down by 45 percent. "A lot of juvenile crime is an impulse thing," says Chief John Kelly. Keeping kids off the streets and away from temptation keeps most of them out of trouble.

Officers take curfew violators to centers, where they talk to social workers while waiting for a parent to take them home. One 14-year-old—caught hanging out with friends at 2 a.m.—told the staffer, "I didn't know anyone cared." Told about a class called "Raising Teenagers," the boy's mother agreed to enroll and make a greater effort to supervise him.

Be specific.

Vague example: Lots of young people stay out at night.

Specific example: Dozens of middle and high school students can be seen outside the 10th Street 7-Eleven after midnight.

Explain why the opposing side's arguments are wrong, or not as strong as your arguments. Avoid name calling.

Example: Critics of the curfew complain that most of teens picked up are Hispanic, and say that Hispanic areas have been targeted for enforcement.

But that's not racism. Most of the complaints about teenagers come from these areas; so do most of the juvenile crime arrests. It's common sense to focus on these neighborhoods. In addition, it's important to remember that the program is not punitive. It's designed to get kids off the streets before they get arrested, and to help parents regain control over their children.

It's OK to admit that the other side has some valid points. Example: Parents Against Racism will monitor the curfew enforcement to see that the laws are applied fairly. That's useful. The curfew must be seen as fair in order to retain community support.

Don't repeat yourself. Don't say everything two or three times because you didn't say it quite right the first time. Repetition turns off readers. So don't repeat yourself.

Avoid telling readers the obvious. Don't waste their time by arguing that education is a good thing or that nuclear destruction is a bad thing. Tell them how to achieve better education or a more secure peace.

Look for clichés in your writing. Eliminate them. (Clichés work as sound bites, not in op-eds.)

Ask someone to read what you've written and tell you if your points and your examples are clear. You probably understand yourself, but the trick is to get other people to understand you.

Here's a useful exercise: Pretend that you have to pay $1 for every word you use. Every time you give readers important information or ideas, you earn $10. Now look at what you've written. Is every word worth $1? Are you losing money? Every word, phrase and sentence should be essential to make your point. If a word isn't doing the job, get rid of it.

Example with useless words: The fact is that Campbell never mentioned Estruth in this debate unless he was asked a question from the panel that required him to do so.

Example with useless words removed: Campbell never mentioned Estruth in this debate unless a question from the panel required it.

Another useful exercise: When you think you've finished, read your article out loud. Listen for clunky wording and over-long sentences. If it doesn't sound right, it won't read well.

Bad beginnings The lead must make a reader want to continue. Here are some beginnings—by professional columnists—that fail to do that (and incidentally, these columns were not published).

> If there is one certainty in our knowledge of the Vietnam War, it is that the arguments among historians—and journalists, politicians and military strategists—will never end. Monuments will rise, memoirs will be written, and statesmen will assert that the lessons have been thankfully absorbed. But this won't settle matters: We don't yet know what those lessons may be, and we haven't yet decided who was right or wrong.

The first paragraph of this op-ed column told us a lot of things we already know. The second paragraph confirmed it. Read between the lines of these two paragraphs and you get the message you don't need to read this.

> If we held an election most people probably would vote spring as their favorite season. For as long as I can remember fall has been mine.
>
> We depend on prospects of crisp October mornings, frost on the vines, the first morning fire of the season, walks through the woods on carpets of new-fallen leaves to sustain us through hot, bug-swatting August days and nights.

It takes a brilliant writer to make a weightless topic worth reading.

I was 1 year old when the U.S. bomber Enola Gay opened its belly high above the Japanese city of Hiroshima and dropped the world's first atomic bomb on Aug. 6, 1945.

This lead should tell us something besides the date of the bombing, which we already know.

All this talk in Washington about a balanced budget is making me uneasy. And it's not because I'm overly concerned about the deficit. Hell, I don't even understand it.

If you haven't researched the topic, it's unlikely readers will follow your argument.

A few days after the Oklahoma City bombing, President Clinton denounced the "loud and angry voices" that poisoned the public debate. "They spread hate," he said. "They leave the impression that violence is acceptable."

Then came columnist Carl Rowan, who blamed the bombing on the "angriest of the angry white men" and even included Senate Majority Leader Bob Dole and House Speaker Newt Gingrich in this accusation. Meanwhile, a smart-alec headline in The New Republic asked, "Did Newt do it?" then added: "Well, no, but he's got some explaining to do."

Clinton's remarks were so poorly crafted that his aides were forced to explain that no, he didn't mean anyone in particular, and no, he wasn't talking about any specific talk-show host—although he later singled out G. Gordon Liddy.

Do not merely summarize what has already been in the paper. If you do need to include background with your opinion piece, do it discreetly, after you have stated your opinion.

————————

It is difficult to imagine that there will be a more defining moment this year in United States history than the bombing of the Federal Office Building in Oklahoma City. As a result of this tragedy, national reactions and public policy debates are now at hand.

There is no opinion in this lead.

————————

An excellent example: We've been critical. We've shown some bad examples. Here is a good one. It was written by a student for our "Everyday Experts" feature. Note its excellent use of personal knowledge about the topic, its no-nonsense use of language, its relentless support of the main point—that private schools are superior to public schools.

For the first time in my life, I am attending a private school. Last year, as a freshman, I attended a public school with a good reputation and students who score high on standardized tests. Unfortunately, those statistics do not reflect the overcrowded classrooms, incompetent teachers surviving because of tenure, and the violence and drugs that abound in the halls.

I received straight A's, but when I had questions I was forced to wait for the students who "need more help because they don't get as good grades as you," as one of my teachers bluntly told me.

Another drawback of public school is the student-teacher program. I understand that people need to learn how to teach, but for one semester I spent English class with a woman who had us read "Of Mice and Men" and other books out loud. I thought that by 9th grade everyone knew how to read by themselves, and English class was supposed to be a time for discussing books.

The violence and drugs that ruled the school were the scariest things. A boy in my Spanish class sold crank on campus. He was caught and suspended for only three days, even though our school supposedly had a "zero tolerance" policy saying anyone involved with drugs or violence is automatically expelled. This year I know of a boy who beat another boy so badly that he went into a coma. The bully was suspended for five days.

Public schools have no control over the kids who are a bad influence, and those who do care to learn get lost in the shuffle.

This year at my private school, which is one-third the size, I have classes of 14 to 17. The students all come prepared for class and ready to learn. The academics are geared toward challenging all students, with help available for those who need it.

Of course there are some drugs and violence on campus, but when the culprits are caught they are expelled.

Private schools don't stifle one's creativity or thirst for knowledge. They provide people with a safe environment where you're held responsible for your actions.

Public schools have become a day care provider for adolescents. Students receive passing grades by merely attending classes.

Getting Your Op-Ed Published Op-eds are significantly more difficult to publish than LTEs because papers are more selective about giving so much space to one article, and therefore have higher standards. Like LTEs, the odds of getting your op-ed placed depend on the size of your newspaper. "I typically read 75 to 100 submissions a day," says Marcia Lythcott, who edits op-eds for the *Chicago Tribune*.[7] Smaller newspapers such as the *Topeka Capital-Journal* and the *Bangor Daily News* receive significantly fewer submissions.

Call the newspaper for which you are interested in writing an op-ed, and ask how they prefer to receive submissions, what the word limit is, to whose attention it should be sent, and if they require any supplemental information beyond your personal contact information. Robert Whitcomb says the best way to send an op-ed is by email, and that you should embed the text of the op-ed in the email rather than include it as an attachment. "The only bad way [to send op-eds] is by fax because they often come in fuzzy." A few days after you submit your op-ed, you should place a follow-up call to the op-ed editor to check on his or her interest in running the article.

Why Do Op-Eds Fail? Many scientists have expressed frustration that their op-eds don't get published. They spend hours drafting and refining their manuscripts, only to have them rejected by one newspaper after another. Richard Gross of the *Baltimore Sun* says there are several basic reasons op-eds aren't published. "Poor writing, too long (beyond 800 words), argument presented poorly, not interesting, boring." Another reason is the recurring problem of inaccessible language. "[Scientists] are writing for a general newspaper audience, not the *Scientific American*. Make it understandable to the reader." William Collins goes even further. "The main reason that op-eds on any subject fail to be published," he says, "is that they are insufficiently targeted. As with any marketing, the product must be aimed at a specific audience and carefully constructed with that audience in mind. For instance, one would never send an article suitable for publication in *Science* to the *Dodge City Daily Globe*, or vice versa. Likewise, a single article is very unlikely to be suitable for both the *Dodge City Daily Globe* and the *Boston Globe*."

PRESS CONFERENCES

A press conference (also known as a news conference) is a staged event where one to four people speak before a group of reporters to either announce or release something newsworthy or respond to something in the news. Press conferences are ubiquitous these days: there are press conferences when a new sports team lands a star free agent, an indicted entertainer proclaims his or her innocence, or a political candidate steps out to make a statement. And, of course, there are scientific press conferences when a team of researchers announces a new discovery.

"A press conference is a very labor intensive event," says Morrow Cater of Cater Communications. "It requires enormous effort creating materials, developing messages, coordinating speakers, and motivating the press to attend." She says engaging the media is one of the most difficult tasks. "Hours upon hours are required to 'pitch' the press, i.e., calling the press multiple times to not just notify but convince them to attend."

Given all the work required, you need to think carefully about whether to organize a press conference. First and foremost, you must have research or views that are important, unusual, or, interesting to editors, reporters, and producers. It helps if your event brings together experts or people that normally don't unite on an issue ("strange bedfellows"), though this isn't required. Ask yourself whether the event will attract television reporters. Print reporters don't necessarily need to leave their office to write a story for a newspaper or magazine. Television reporters, on the other hand, invariably need video footage. (If there is some doubt that reporters will attend your press conference, it can be just as effective to email or mail your materials to reporters and speak to them over the phone.)

"Your story must have compelling pictures, or a way to tell the story visually, in order to make it into broadcast news," says Cater. "Some people make the mistake of thinking that a press conference solves the problem by providing a picture for the cameras. That is true only to a point—you must have real news to attract press to a press conference."

Just like a press release, a press conference is a venue in which you deliver your main messages and talking points in a quotable form so reporters can produce stories. Be concise and make your research or views compelling and clear, sprinkling your remarks with sound bites as you deliver your messages. Never have more than four people speak at a press conference. If there are only two speakers, they should speak for no more than five or six minutes each. If there are three to four speakers, they should speak for three to four minutes each. Reporters should be invited to ask questions as early on in the press conference as possible.

Preparing for a Press Conference In advance of the event, create your compass of main messages and talking points. If more than one person will be speaking, each person should choose one or two of the messages as their main focus of discussion. Brainstorm all the possible questions reporters may ask you, and have your answers ready, always aiming to bridge back to your talking points. If possible, try to schedule your press conference at 10 a.m. or 11 a.m. That will give newspaper and television reporters plenty of time to put their stories together for the evening news or the next day's paper.

Hold your press conference in a place that helps tell your story. For example, if the subject of your news event is sea-level rise, hold the event at a dock; if it's related to a policy issue, hold the event in front of a government building (you might need permission or a permit first); if you are releasing university-sponsored research, hold the event in a university building. Just make sure the event is located in a convenient place for reporters to attend. In Washington, DC, many private and nonprofit groups stage their press events at the National Press Club because reporters work in the building or nearby. Hotel conference rooms are also a good option if there are no other centrally located venues. If you expect a large turnout of television reporters with cameras, you should secure loan of a "mult box," a device that connects to the podium microphone (via a long cable) and allows reporters to plug in audio recorders away from the stage.

Once you have reserved a space and chosen a time for your press conference, send a press advisory to reporters a week before the event, then resend it a day or two before the event. The advisory delivers all the logistical details of the press conference and a very short teaser of why the event will be newsworthy (without giving away the news). Be sure to send the press advisory to your local or regional Associated Press and Reuters bureaus and request that it be added to their daybooks (shared calendars for other media in your area). You or your institution's media staff should then call key reporters you would like to attend the press conference to make sure they received the press advisory and encourage them to attend.

Press Conference Day. On the day of your press conference, you (or your organization's media staff, if applicable) should arrive an hour early to check that the room is properly arranged. Also bring a sign-in sheet to collect the reporters' names as they arrive.

An emcee (someone affiliated with your institution, or the first speaker if he or she is comfortable with the extra role) should briefly introduce the speakers and summarize the importance of the event. Each speaker should then take a turn communicating his or her main message, along with background information and talking points. "Always speak with the knowledge that in the best of all possible television worlds, you will get 10-15 seconds on screen," says Cater. "Keep your sentences short and punchy, keep your thoughts cogent and concise, keep your language jargon-free. You must speak for the lay audience at all times."

After the last speaker, the emcee should open the floor to journalists' questions. Ask the reporters to identify themselves and their media outlet when they have a question. Reporters should be allowed follow-up questions, but don't let one reporter dominate the questions. Make sure all reporters have an opportunity to pose questions to the speakers.

Finally, never let a press conference run over an hour. Reporters are busy people, and if you talk too long, it means you didn't do a good job of concisely explaining your research. Most press conferences can be completed in thirty to forty-five minutes.

TELEPHONE PRESS CONFERENCES

If you have a story that will interest large numbers of print reporters, but not television reporters, you might consider holding a press conference over the telephone. Telephone press conferences are a convenient way to talk to many reporters at once, especially if you are releasing a report that is regional or national in scope and the reporters are in different media markets. A telephone press conference is promoted and presented much like an in-person press conference, from the press advisory to the allotted time for speakers. However, phone press conferences can lack the energy of a live event so short presentations and responses are key.

To prepare for your call, contact a conference call company a week before the event to receive a phone number and access code for reporters. On the day of the event, when the reporters first call in, make sure they can only listen to the speakers; their voices can be muted until the time when the press conference is open for questions. Reporters who wish to ask questions can press a button on the phone to be placed into a queue.

VISUALS

Print publications like to illustrate stories, so you should look for opportunities to present your data and findings with interesting graphics or visuals. If you can present statistical information in a chart or other form that is easy to understand, share it with reporters at your press conference, during your cultivation meetings, or even over email. Likewise, let them know if you have interesting photographs or can take the reporters to see something with their own eyes.

If you have video images of your research, or areas affected or aided by your research, provide this footage to television reporters so they can use the images in a story (this footage is called B-roll, as often the video will be supplied on beta tape). Radio reporters might also be interested in B-roll, especially if it has audio they can use in their stories. They might even be able to post it on the station's website.

PRESS KITS

Press kits (also known as media kits) are sets of materials sent to a reporter either by mail, email, or in person. A press kit can include all of the following:

- reports
- press releases
- background papers
- fact sheets
- experts' biographies
- contact information for other experts
- visuals (charts, graphs, and photographs)
- B-roll
- quotes from people who endorse your work or views

Whether you are releasing a major report or simply getting together with a reporter for a cultivation meeting, you should consider producing a press kit. Never send a press kit via email (also known as an electronic press kit) without first asking the reporter whether he or she is interested in the work. Otherwise, the reporter might delete the email before opening it given concerns about unknown senders and email viruses.

GET HEARD ON RADIO

Communicating over the radio should be a central component of your work with the media. "The scientific community must stay connected to radio and their audiences, respectively," says Aric Caplan of Caplan Communications. Caplan, a former executive producer for a morning radio show in Pittsburgh and a program director for a talk radio station in South Florida, notes, "Radio has preserved its value due to the advent of satellite radio and the popularity of Internet radio. Consequently, audiences seeking news and information are on the increase. Plus, factor in loyal listener involvement in public radio."

Public radio and commercial news radio are both worth pursuing because they offer a mix of news programs and discussion formats (i.e., talk radio). Send your press releases to radio stations in your area—both public and commercial—and call to follow-up. You may need to send the release to the different producers who are in charge of the various radio shows, in addition to the news department.

If given a choice between radio stations, we suggest contacting your local NPR station first. NPR is unique in that it often puts together longer pieces than commercial talk radio. Richard Harris, a longtime science reporter for NPR, says, "I like to get people to tell me a story." Like most news outlets, NPR is looking for stories that are new or interesting, but it will often go more in-depth into a science issue compared with commercial radio. Many of NPR's science stories, like those of other media outlets, are generated from *Science, Nature,* and other science journals. "Scientists know that if they have something really hot, they go first to *Science* and *Nature,*" Harris says. "So when we're looking for the hottest stories we look first in *Science* and *Nature,* and [also] medical journals." Harris says that with a beat as large as his it is hard to stay on top of the smaller journals and research coming out of other universities, but he still wants to hear from scientists when they have news.

Radio is a great way to communicate to the public because it provides the fastest turnaround of breaking news. And unlike newspapers, you can reach a diverse audience. "One of the most remarkable strengths of radio today is its ability to reach more minorities and underserved communities," says Caplan. "African-American and Latino audiences comprise the greatest growth in radio listenership."

If you have a story with broader national implications, your media staff or a public relations agency can set up a "radio tour." A radio tour is a series of interviews you conduct from your office or home, and each interview is aired on a different radio news program around the country. It is a very effective way to reach a wide audience in a short time.

SATELLITE TELEVISION TOURS

If you are looking to communicate your research on a major study or issue, and have a large budget, you may want to consider a satellite television tour to promote your messages. Satellite television tours are similar in structure to radio tours. For example, if you are a scientist at Boston College, you can go to a studio in Boston and conduct interviews that appear, via satellite, on news programs in Memphis, Indianapolis, and other cities around the country.

There are several benefits to satellite interviews. "By conducting a TV satellite tour, spokespeople and their business save the costs of flying and lodging, time away from their office, and exhausting themselves touring across the country," says Caplan. In addition, he says, "During one three-hour morning a satellite TV tour would enable representatives to reach audiences of between 15 and 22 local NBC, ABC, CBS, and Fox affiliates. Interviews are conducted on the major local news programs, which lead into the *Today Show*, *Good Morning America*, and the *Early Show*. Satellite TV tours also add prestige to your organization by appearing on local newscasts where membership and sponsors may be located."

As long as you have a newsworthy topic, provide a good local or regional story angle (showing why people in that community should care), and have some visuals, many local television stations will be interested. Stories relating to health, safety, and the economy are often the most attractive to these reporters.

You should only consider scheduling a satellite media tour once or twice a year. "Local NBC, ABC, CBS, and Fox affiliates will not justify participating frequently with one organization or university repeatedly," says Caplan. "And, the news any academic or public policy group has to convey may not be dominant enough to merit a full TV satellite tour more than twice a year."

Satellite media tours are not cheap. In New York City and Washington, DC, production agencies charge up to $20,000 for a satellite tour. A second camera (for taping additional speakers) costs an additional $1,500. These prices include studio time, production staff, makeup, satellite airtime, and transmitting the story to the satellite. Agencies should provide references and examples of previous satellite media tours.

VIDEO NEWS RELEASES

A video news release (VNR) is like a press release, except the information is presented in a visual form. A VNR features sound bites from an expert or experts, video footage that brings the subject of the issue to life, as well as other facts that convey key messages and talking points. Paul Frick, a partner at Home Front Communications, which produces and distributes VNRs for its clients, says a video news release should tell a compelling story and provide background information that will "answer many of the questions about the story a producer/reporter might have."

Businesses, universities, trade and nonprofit organizations, and the government create VNRs as a way of pitching news stories to local and national television news programs. Television stations frequently use VNR footage or information in their news stories. Sometimes they will even air pre-edited VNR story packages that appear to viewers to be put together by the station itself. It's best, though, to produce VNRs that include "slates"— text of your messages and key talking points—along with your videotaped sound bites and B-roll. That way local reporters will put together stories based on the information you send, as they do with normal press releases.

VNRs can be an effective way to reach large audiences around the country, as television is the largest single source of news for most Americans. "A typical story airing on CNN Headline News may reach 200,000 to 250,000 viewers depending on the time of day," says Frick. "If the same story were to air on one station in Philadelphia it would generate an audience twice as large. So there is no better medium for reaching large audiences than local television, which makes video news releases very effective at reaching the general public with important information."

However, there is a trade-off.

Local TV news is produced for viewers looking for as much news as possible in the shortest amount of time. Because TV news stories are so short (usually a minute and thirty seconds), there is little or no time for scientific detail, so VNRs must tell a story with information, sound bites, and images almost anyone would understand.

It's probably not often that you have a story worthy of a VNR. Health and consumer stories have the best odds of seeing air time. "Stories of interest only in the scientific community will probably not fare well in the television news environment," says Frick. But if they are on topics that could be of interest to a wide audience and can be presented in a way that highlights their value to individuals' day-to-day lives, then it would be appropriate to consider using a video news release as a means of getting the news out.

Like satellite television tours, VNRs are expensive, costing between $23,000 and $35,000 for the full range of production and distribution services. That includes advice on how to tell the story, coordination of camera crews, pitching the story to reporters, and distributing the VNR through satellite feeds and overnight mail. The service should also include an analysis of media results, including the number of stories the VNR generated, the size of the viewing audience, and some sample stories.

You can save money on VNR production costs if your university or organization has access to equipment and production experts who can videotape you and help compile B-roll images. Your PIO or media staff can pitch the VNR to local television stations to further reduce costs. However, if you have the budget, it might make sense to work with professional firms, because they know the kind of stories that interest local television stations and have the local contacts to help ensure they are aired.

CHAPTER 8

The Scientist as Celebrity and Activist

WHEN PHYSICIST Len Fisher first took a call from an advertising agency, he had no idea this contact would soon give him greater insight into how to improve public understanding and appreciation of science, a problem that had been on his mind for years. Fisher, who has been said to "put the fizz in physics," is a member of an unusual breed of scientists who have been called science popularizers, citizen scientists, public scientists, and visible scientists—researchers not content just to conduct science or even just publicize their own research.[1] They are the ambassadors and sages who are the public face of scientific knowledge and research, sometimes more well known for their television show appearances, best-selling books, congressional testimony, or social or political commentary than their scientific accomplishments.

Neal Lane, the former administrator of the NSF and science advisor to President Clinton, has become an apostle for this public-spirited class of scientists whom he calls "civic scientists." These scientists and engineers, he explains, "step beyond their campuses, laboratories, and institutes into the center of their communities to engage in active dialogue with their fellow citizens."[2] A physicist and professor at Rice University, Lane says that where public appearances were once discouraged by scientists' peers, today they are accepted (if not applauded).[3] He urges scientists with communication skills to use them: "When you see an opportunity, answer the call for God's sake and spend the time to give the interview or go to Washington to testify."[3] As we'll see below, these activities are not without risk, though even many scientists who got in hot water say they did the right thing.

Scientists can engage in public dialogue through a broad spectrum of activities. One way of thinking about these activities is to visualize them as being arranged in a spectrum of least to most political. At the apolitical end are activities that improve public appreciation and understanding of science and scientific topics. At the other end are overt attempts to influence governmental policies.

POPULAR SCIENCE OR SHAMELESS OPPORTUNISM?

In 1998, Len Fisher was a physicist at the University of Bristol in England who had previously gained brief notoriety by publishing a tongue-in-cheek paper on the "physics" of sex.[4] His opportunity to reach out to the public came when he was tapped by an advertising executive representing the McVitie's line of "digestive biscuits" (or cookies). The agency wanted a scientific angle for a campaign to pump up sales of the snacks for its client, the United Biscuit Company. The manufacturer brags that fifty-two of its Chocolate Digestive biscuits are consumed every second in the United Kingdom. McVitie's was planning a "National Biscuit Dunking Week" and wanted Fisher's assistance in anchoring this publicity stunt on the firm footing of the "science" of dunking.

Fisher agreed to help, and began conducting some laborious investigations with colleagues over tea and cookies. In fact, the actual subject of the experiments *was* tea and cookies. Baked biscuit dough is a composite of starch granules glued together with sugar and fat. When dunked in hot tea or coffee, the voids between starch grains fill with the liquid in a process called capillary action, the same way blotter paper soaks up ink. According to research in Switzerland, dunking improves the flavor of the biscuit by as much as a factor of ten, but—here's the rub—the stiff, baked starch becomes pliable when soaked. The hot drink melts the fat and dissolves the sugar, which is like stripping mortar from a brick wall. In almost no time, the soggy mass can no longer support its own weight and splashes down unceremoniously into the cup—a mishap that biscuit makers claim happens to one of every five biscuits.

Fisher decided that the amount of time a wet biscuit remains intact is too brief, a problem he set out to solve. After examining the causes of biscuit failure, extending the life of the soggy snack was easy, though predicting how long it would take the average biscuit to collapse required further theoretical analysis. Fisher discovered that the most common method of dunking, dipping a biscuit vertically into a hot tea or coffee, perpendicular to the drink's surface, is unsound because the treat becomes soggy and structurally weakened in a matter of seconds. However, if the biscuit is dunked horizontally, with only the bottom surface in contact with the liquid (a process that involves holding the sweet like a Frisbee ready to be thrown), Fisher says it stays intact four times longer. The reason for this remarkable increase is because the biscuit's dry side can bear the weight of the soggy side as long as the liquid hasn't soaked all the way through.

When Fisher made his "research" public, both print and broadcast media in the English-speaking world embraced it enthusiastically. Dozens of newspapers, including every major British daily, ran stories on the topic. Fisher continued to field calls from as far off as Australia and South Africa for months. (Notably, not one major paper in the United States reported on Fisher's publicity stunt, possibly because the spongy donut, America's favorite dunking sweet, holds up better than the brittle biscuit when saturated with a hot drink). "Even the Nobel prizes don't receive such coverage," Fisher exclaimed in a *Nature* commentary he wrote on his experience.[5] A year after the dunking stunt, Fisher was awarded an "Ig Nobel Prize," given each year for research that "first makes people LAUGH, then makes them THINK."[6]

What accounts for the media's notable attraction to Fisher's biscuit investigations? And what lessons does his experience hold for other scientists hoping to increase public understanding of science or to improve the empirical and theoretical foundation of public policies? Fisher says his experience with biscuits demonstrates that though the media may not be leery of all technical topics they are most comfortable communicating the "science of the familiar" (though he notes that they are also fascinated with explanations for the "grand questions," like the fate of the universe). To his surprise, the press was attracted by his use of a formula, the Washburn equation, to describe the behavior of a soggy biscuit, which suggests to Fisher that equations simple enough to be described briefly but complicated enough not to appear obvious will pique the interest of print and television reporters alike.

Fisher says he has learned how to avoid a problem common to many researchers who try to communicate with the public. He says they fall short due to a proclivity for one of the two poles of the sophistication spectrum: they are either oblivious to the low technical competence of their audience, which will drown in "excruciating detail," or they pare back a topic's detail so much that it becomes sterile. He says he's learned how to tailor these details to the needs and interests of the audience. He has come to believe that people want details about *why* researchers are asking the questions they do, and are less interested in *how* they get the answers (e.g., the methodology). For instance, many people will probably be interested in an astronomer's research on the luminosity of distant stars if they learn that this could help to settle whether the universe is going to keep expanding forever or eventually contract in a "big crunch." They will be considerably less interested in the details of the detector or how the researcher corrected for errors. "You don't have to be a writer to enjoy something well-written," the physicist explains, "or an artist to enjoy something well-painted."

Since his debut dunking biscuits as a scientist for the common man, Fisher has written two books about the science of the familiar. Asked if he regrets any inches added to the British waistline on his account, he says he has "no qualms about sugar in the diet." He adds that he gets about one email a week from advertisers asking for a scientific take on one product or another to produce a campaign like the dunking stunt. He has rejected all but a small handful, those that "I could use to make science accessible."

TAKING HEAT FOR MAKING SCIENCE COOL

With his intriguing investigations into the activities of everyday life, Fisher joins a distinguished fraternity of public scientists that includes Barry Commoner, Jared Diamond, Sylvia Earle, Paul Ehrlich, and E. O. Wilson. These are some of the most famous of the hundreds of scientist foot soldiers who help improve public understanding and appreciation of science and help give public policies firm, scientifically sound foundations.

A generation ago such scientists were often looked at askance. The quality of their scientific achievements was often mocked behind their backs and their motives were called into question. When Carl Sagan, the consummate celebrity scientist, was a graduate student who managed to get himself quoted in the *New York Times*, a senior scientist in his field sarcastically remarked, "I've been following your career in the *New York Times*," which the young scientist took to mean that he had "done something bad by being quoted."[7] Beginning in the 1970s, Sagan became an omnipresent spokesperson for science with best-selling books like *Cosmos*, a popular television series on astronomy and the search for intelligent life beyond Earth, and regular appearances on the *Tonight* show with Johnny Carson.[8]

It was no secret that many scientists found the astronomer's wide-ranging public activities uncouth; in fact, this professional displeasure was so pronounced that scientists still speak of the "Sagan effect" to describe a backlash against researchers who garner extensive public exposure.

Biologist Thomas Lovejoy, president of the H. John Heinz III Center for Science, Economics and the Environment, recalls that when he was a graduate student in the 1960s researchers who took their scientific expertise outside the lab "were regarded with suspicion." Today, he says, "it's seen as something everybody's got to do." Physicist David Grinspoon attributes at least part of this change of heart to pangs of "collective guilt" for the shabby way the scientific community, especially astronomers, treated Sagan. Grinspoon, a planetary scientist and author of two nontechnical astronomy books, knew Sagan as a mentor and family friend. He says that since the well-known astronomer's death in 1996 the scientific community has come to appreciate the debt it owes the telegenic researcher. Sagan stimulated widespread interest in the search for extraterrestrial life, space travel, and planetary studies, which has helped sustain funding for these activities. "There is an impulse to look down on them," says Grinspoon, referring to science popularizers, "then they remember Carl."

As we'll see below from Grinspoon's own experience, while tolerance for such visible scientists has improved, they still sometimes face disapproval or worse. Len Fisher says when he began publicizing the science of the familiar "there were certainly people who said, 'why are you wasting your time?' 'why not do something serious?'" He still hears occasional "snide remarks," but as colleagues have begun to see the value of his work, he has received more appreciation. He doesn't recommend that young scientists follow his footsteps (he began in his mid-50s) because "if it backfires it can kill your career." Still, like many scientists who use their specialized knowledge in a public way, he says he has few regrets. In March 2004, Neal Lane advised students and faculty of his alma mater, the University of Oklahoma, that in a time when some public policies at the national level are being made in the absence, or even in defiance, of empirical or theoretical justification, it is critical for scientists to be heard in public. "The risk of not speaking up," he said, "is that if people are allowed to let special interests, narrow ideologies, or political agendas trump the truth in science, then two things will happen: we will get bad policies (even dangerous ones); and the public will begin to lose trust in the value of science and the government's ability to use science for the public good."[9]

David Grinspoon's career as an author of popular science books shows that scientists can't engage in this kind of activity without occasionally looking over their shoulders, especially if they are at the beginning of their careers. In 1999, Grinspoon was denied tenure at the University of Colorado at Boulder at least in part because he had written a book for lay readers, a step he advises young faculty members to weigh carefully. Between 1990 and 1994, the Colorado professor was a member of the NASA team interpreting radar imaging from Magellan, a spacecraft orbiting Venus. In many ways, such as size, mass, and distance from the sun, Venus is the most Earth-like planet in the solar system, though its surface temperature (400 degrees Celsius) is hot enough to melt lead. Before Magellan's mission, no one knew what the surface of Venus looked like due to the dense clouds that perpetually shroud it. The space probe used radar to see through the clouds and sent back enough data to create a detailed, nearly complete map of the planet's surface. Grinspoon's 1997 book *Venus Revealed* was the first to explain what the team found on the mysterious planet.[10]

Grinspoon wrote *Venus Revealed* for the educated public and included his personal musings about how the research team's discoveries might help make Earth a better place to live. That might seem harmless enough, yet two years later his university denied him tenure. It wasn't that he didn't have the support of his colleagues. The faculty of his own Astrophysical and Planetary Sciences department voted by a wide margin to grant him tenure based on his record as a teacher and researcher and on his "service" to the broader community. However, the dean of the College of Arts and Sciences (which includes Astrophysical and Planetary Sciences) recommended that the university deny the junior faculty member tenure. In his official letter on Grinspoon's case, the dean said that the young scientist "appears to have dedicated much of his time to the popular (in both senses) book *Venus Revealed*. His research productivity seems to have been adversely affected."[11] Grinspoon believes that the issue wasn't really the quality of his research but "prejudice against junior faculty writing to a mass audience."

The university's decision did not stop Grinspoon from courting further controversy. He has since written a second book for lay audiences, *Lonely Planets*, an extended essay on the possibility that life, intelligent or otherwise, exists beyond Earth.[12] The book mainly concerns scientific research, but Grinspoon spends a full chapter discussing UFO sightings and accusing some of the scientific community's self-styled "debunkers" of UFOs as sometimes being as immune to facts and rational analysis as the true believers they ridicule. In their attempts to counter what they fear could be a tidal wave of ignorance and irrationality if false claims go unanswered, Grinspoon charges the skeptics with treating all UFO reports as equally invalid. Though he finds no evidence that extraterrestrials have visited Earth, he believes some sightings have never been adequately explained. Grinspoon's website, www.funkyscience.net, extends his arguments and also offers a glimpse of his sense of humor. For example, the scientist admits he "played lead guitar for a band called the Geeks years before being a geek became cool."[13]

Though unconventional, Grinspoon is no crank. He is now the Curator of Astrobiology at the Denver Museum of Nature and

Science as well as a principle investigator for several NASA projects. He says the primary objective of his writing is to "spread the joy and wonder of science" beyond "the priesthood with the arcane knowledge and language." He says his books, though not aimed at scientists, also serve a function in advancing science. They allow him to make speculations that, however well informed, would not be possible in peer-reviewed publications. He can take ideas that "are too out there," such as the potential for life in the clouds of Venus, and "put them up the flag pole and see if they salute." This particular idea, which he described in *Venus Revealed*, was later referenced in scholarly papers. He also claims to have what he jokingly calls a "subversive agenda": promoting a global perspective on our planet, which he believes "is necessary for long-term human survival." In the final chapter of *Lonely Planets*, for instance, he concludes that if there is intelligent life on other worlds, it has probably passed through the same "bottleneck" humans face today—having enough technology to destroy itself but not enough wisdom to ensure that it doesn't. Which leads him to the first question he would ask an extraterrestrial "after the usual pleasantries:" "Not 'How do you build your wonderful machines?' but 'How did you learn to live with yourselves?'" Grinspoon says he gets "a lot of positive feedback from people who have read my books," including his astronomy colleagues. But based on his "one negative experience," he still tells untenured professors who want to write for the public, "watch it."

BALANCING RESEARCH AND PRINCIPLES

One way to be a public advocate for science, especially for those shy of the political arena, is to beguile the science-averse public with alluring tidbits of knowledge, as Len Fisher and David Grinspoon do in their books and interviews. Those who are more comfortable in the spotlight might want to actively campaign for specific public policies or encourage the public to view new policies, reports, and statements about their significance, with a healthy dose of skepticism. Some scientists conduct these activities in the free moments of an otherwise traditional career, but others make such activity a central part of their life and work.

Jonathan Beckwith has spent more than thirty years and a good deal of his waking hours advocating such skepticism, especially in fields associated with his specialty, microbiology. The Harvard Medical School professor argued in his 2002 memoir, *Making Genes, Making Waves*, that researchers can do good science and be social activists at the same time (with his own career serving as Exhibit A).[14]

As a young scientist, Beckwith led the first team to isolate a single gene in a test tube. This achievement was an important milestone in the development of techniques for manipulating genes that, along with discoveries by many others, launched the revolution in genetic sciences in the 1970s and 1980s. The research earned him the prestigious Eli Lilly Award of the American Society of Microbiology, given to young investigators who make outstanding contributions to the science. Beckwith's current research involves using bacterial genetics techniques to investigate protein folding in cells, cell secretion of proteins, the structure of cell membranes, and cell division. He is a fellow of the American Association for the Advancement of Science and a member of the National Academy of Sciences, two high honors that testify to his standing as a scientist.

Beckwith has risked that standing through his many activist activities. He was a longstanding and active member of Science for the People (until it disbanded in 1993), a group of radical scientists formed in 1969 to oppose what its members considered misuses of science, such as research comparing the intelligence of whites and blacks (which they branded racist) and research to improve the "electronic battlefield," an attempt to use technology to make U.S. soldiers in Vietnam more lethal. (In the interest of full disclosure, we mention that Grossman was a member of Science for the People in the 1980s, where he met Beckwith.) When Beckwith and his colleagues' genetics breakthrough was announced in 1969, he simultaneously launched what has turned into a lifelong campaign to raise public awareness about problems that can accompany research results. Along with two colleagues, he called a press conference to point out the possibility (since realized) that the ability to purify genes could pose dangers to humanity by giving scientists the power to alter human genes. The scientists explained that although this power might have important health benefits, it could also be used for nefarious purposes, such as to abet discrimination. Their warnings gained widespread press attention with headlines such as "The Gene Isolated—for Good or Evil?" in the *New York Times* and "Scientists Isolate Pure Gene from Bacteria with Virus: Test-Tube Man Feared" in the *Los Angeles Times.*[15]

In a 1970 speech Beckwith gave after receiving the Eli Lilly Award, he denounced practices that encourage the overuse of antibiotics in the drug industry. Though he didn't mention Eli Lilly by name, the relationship between his remarks and the award's sponsor, a major drug maker, was obvious. In protest of pharmaceutical industry practices, Beckwith declined to accept for his own use the Eli Lilly prize money. Instead, he donated it to the militant Black Panthers organization, which was then challenging persecution by New York City police.

As described in his frank memoir, much of Beckwith's activism has been an expression of his concern over the social impact of genetics research. The microbiologist was deeply affected by learning about past abuses of genetics research, such as the eugenics movement in the United States early in the twentieth century, which said protecting America required discouraging people of "inferior" genetic stock from procreating. State and federal laws influenced by eugenics included mandatory sterilization of women of low intelligence or with certain behavioral "abnormalities" (such as a history of criminal activities), prohibitions on interracial marriage, and restrictions on the immigration of Jews and other groups considered "undesirable." Germany's "racial hygiene" movement, which culminated in the Nazis' genocidal plan to exterminate Jews, was energized by eugenics in the United States. Although the eugenics movement died out by World War II, Beckwith has accused some contemporary genetics scientists of using unsound science to justify social policy prescriptions or draw unfounded conclusions about the biological foundation of social arrangements within society.

For instance, a 1965 study of Scottish prison inmates concluded that men with an extra Y chromosome ("XYY men") were predisposed to "unusually aggressive behavior."[16] The research was later discredited as sloppy, but at the time it received widespread attention in the media, especially when a famous mass murderer incarcerated in the United States was identified—incorrectly—as having the extra Y chromosome. In 1973, Beckwith discovered that some of his medical school colleagues were conducting a long-term study of children with an extra Y chromosome. After getting some information about the research, Beckwith concluded that the parents had been recruited at an emotionally vulnerable moment (when the mothers were about to go into labor) and had signed misleading consent forms (which, among other improprieties, didn't specify the true focus of the research). He considered the study design to be so poor that it would render the results useless, a failing that called into question the wisdom of subjecting children to the risk of being stigmatized by the research. Beckwith also worried that if the parents became aware of a supposed genetic predisposition for aggression in their children, the study could influence child rearing and unintentionally *cause* the hypothesized behavior. He therefore filed an official complaint with the school's Committee of Inquiry, a process that culminated one year later in a bitter meeting of the entire Harvard Medical School faculty. Beckwith's challenge was rebuffed, but the study's researchers ended up bowing to the public pressure sparked by their colleague's protest and suspended the work.

This scientist's most enduring legacy may be his repeated efforts to knock science and its practitioners off the pedestal of objectivity, which he believes needlessly insulates scientists from public scrutiny. He has repeatedly questioned the idea that scientific research is value-free by pointing out the importance of factors such as luck, intuition, and personal idiosyncrasy. The impersonal form and passive voice of scientific journal articles, Beckwith believes, perpetuate the "myth of pure objective science" by distilling out any hint of the subjective human imprint on scientific discovery. These articles remove the bends, twists, and blind alleys of the research process so that the final description of an experiment resembles the actual tortuous path from hypothesis to result no more than a superhighway resembles the country road it replaces. This misrepresentation of reality led Beckwith to defy the traditional style of paper writing in a 2000 article published in the *Proceedings of the National Academy of Science*.[17] The article, which began uncharacteristically, "This is the story . . ." described the ten-year meandering path that led to his findings about the protein membranes that make up a cell's walls. Beckwith also notes that as human beings, it is impossible for scientists to be entirely objective, especially about the human race. He therefore cautions biologists who attempt to explain human social arrangements or behaviors by drawing on the principles of evolution, such as evolutionary explanations for rape, that they will inevitably bring preconceptions to their research that bias their investigations and conclusions.

Beckwith owes his deep involvement in the politics of science to the diverse currents of politics and ideology he has lived through—the counterculture revolution, the opposition to the Vietnam War, and the initial willingness of prominent microbiologists to raise public concerns about the possible adverse consequences of genetic engineering. He says he was fortunate to work in an academic setting that, up to a point, tolerated unconventional activities like his involvement in Science for the People. And although he wants to encourage, by his own example, today's young scientists to combine science and activism, he admits that the combination of historically rare factors that propelled his career make this trajectory hard to replicate today. While Harvard "has a tradition of attracting progressive people," he says, "That isn't true everywhere." Beckwith worries that there may not be a new generation of young scientist-activists ready to fill the shoes of his generation. "The situation today is a little scary," he states. "There are not younger people coming up. I don't see them doing this thing."

Looking back at his intertwined careers as both a scientist and activist, Beckwith says the activism has "added something to my life." And more importantly, he believes his work has contributed to "an awareness that ethical and social issues have to be considered" in genetics research. For instance, within the initial budget for the Human Genome Project, the massive crash program to identify the 32,000 human genes and transcribe the entire sequence of DNA code (made up of three billion elements), 5 percent of the research dollars was allocated to study the accompanying ethical, legal, and social issues. Beckwith was one of the first members of an advisory group formed by the U.S. Department of Energy and the National Institutes of Health to oversee this research.

The Harvard researcher is quick to note that stepping from the lab into the street and substituting his pipettes for petitions has not always been easy. He also acknowledges that his activism has hurt his research in a number of significant ways. Between 1970 and 1984, for example, he won no scientific award or honor—a dry spell he concedes was caused in part by insufficient research progress during that period. He says at least one significant honor, membership in the National Academy of Sciences, was delayed for years in reaction to his 1970 Eli Lilly speech. The substantial time Beckwith devoted to the affairs of Science for the People in the mid-1970s distracted him from his lab, causing him to miss at least one "fairly important" discovery. He also barely avoided having a formal request to remove his tenure filed by a group of faculty members who were reportedly upset that he had brought public attention to the XYY issue. That plan was shelved, he says, when the medical school rejected his request that the research be halted.

THE PATH TO BECOMING A POLITICAL SCIENTIST

Stephen Schneider, a Stanford University professor, has also spent much of his career focused on how scientific research is used to make government policies. Since 1970 he has studied and written about how heat-trapping carbon dioxide emissions are altering the climate. He has co-authored hundreds of papers on the subject, written or edited half a dozen books, and edits the peer-reviewed journal *Climate Change*. He also says he devotes at least one-third of his waking hours communicating with the media about the politics and science of climate change. "If you want anything to happen, you have to get it out there," says the media-savvy scientist, explaining why he spends so much time talking to reporters on the phone and in front of television cameras. "You have to tell the world what is going on." And tell he does. The LexisNexis database lists more than 300 newspaper and magazine articles in which his name appears in the space of a decade—that's two and a half articles a week (and the database doesn't include coverage in smaller newspapers, local television stations, local radio, and most international media). For his contributions to research on and communication about climate science, Schneider received a MacArthur Fellowship in 1992.

Why is he so driven? Because, says the Stanford scientist, "the world is not going to beat a path to your door just because you have good science." Just a generation ago, before scientists had solid evidence that humans were warming Earth to higher temperatures more quickly than at any time in thousands of years, climate science was a genteel discipline with relatively little federal funding and almost no public profile. Debates about research results were carried out quietly and generally politely at conferences and in academic journals. Today these same researchers are at the heart of a bitter political battle that could determine the fate of entire ecosystems, low-lying cities like New Orleans, and billions of investment dollars. The fossil fuel industries, which are concerned that their short-term profits could suffer if the United States actively sought to reduce heat-trapping emissions, and their lobbyists pay a handful of researchers to sow doubts about climate change. These professional doubters vociferously poke holes in research that supports the worldwide scientific consensus that humans are causing warming. As a result, new scientific papers are sometimes the subject of bruising public battles fought on television, in newspapers, and in congressional committees. Schneider says if he and like-minded colleagues don't get involved, the fossil-fuel-funded scientists could win the day regardless of the merits or veracity of their arguments.

As a prominent and outspoken representative of the scientific consensus, Schneider has been at the center of many of the recent debates connected to global warming. In the process he says he has been "dumbed down," subjected to "character assassination," has had quotes trimmed and twisted so that the meaning was turned "upside down," and has had fabricated quotes as long as an entire paragraph falsely attributed to him. But the Stanford professor perseveres because, as he puts it, he has no choice. "In my view, staying out of the fray is not taking the 'high ground,'" he says, "it is just passing the buck."

Schneider has advice for scientists who want to join him in stepping outside the lab into the public arena, especially if they plan to make policy recommendations in fields that, like global warming, have become highly polarized by politicians. To his students who ask about media appearances, he says, "Do it a lot or not at all." Like a financial counselor who advises investors to buy diverse portfolios of stocks to balance losses in some securities with gains in others (presumably yielding net gains), Schneider says repeated media appearances will help reduce the impact of a single lousy story. "Some stories will make you look foolish to your colleagues while others will make you look better than you deserve," he told an audience at the 1996 meeting of the American Association for the Advancement of Science.[18] (Schneider acknowledges, nonetheless, that scientists not willing to make the kind of commitment he has made to political and media visibility can still be involved in many of the ways discussed in this book, such as writing op-ed articles and letters to the editor and giving talks at small community meetings.)

As we discussed in chapter 5, media appearances require serious preparation. We're not talking about knowing your subject matter—that goes without saying. We're talking about how to convey your message. Schneider recommends that scientists who are going to discuss policy matters with the media try as hard as possible to make the distinction between what *is* and what they think *should be*. In other words, they must endeavor to prevent their opinion of what *should be* from influencing the many subjective judgments they make about what *is*. That doesn't mean scientists need to limit themselves to stating scientific facts ("I, too, am a citizen entitled to preferences," Schneider says), but it does mean scientists must be clear about when they are educating and when they are advising.

Researchers who achieve the near-Zen-like state of understanding their own prejudices and the weak points of their field's scientific knowledge are not yet ready to go public. Schneider says these scientists still face a "double ethical bind" which they must confront, if not overcome. On the one hand, their professional duty requires them to explain their science accurately. This involves describing the full range of outcomes and specifying the likelihood that each might occur. In the case of the greenhouse effect, for instance, the spectrum of outcomes ranges from minimal transformation of the climate to an abrupt climate catastrophe that breaks up ice caps, raises sea levels, and inundates coastal cities. Neither of these outcomes is very probable, of course. More likely are more moderate (though still very serious) scenarios that include a steady rise of sea level by tens of centimeters and of global temperatures by several degrees centigrade (changes which, though gradual, are likely to eventually exact huge health, economic, and environmental costs). Schneider says a scrupulously honest scientific presentation would also require a catalog of the uncertainties involved in calculating outcomes.

The "other hand" of the ethical bind is the fact that scientists trying to influence policy want their contributions to be effective. But, in a complex field like climate change, the text of a report containing the sort of thorough analysis described above might contain more pages than the New York City white pages. Indeed, oral testimony containing all the elements Schneider contends would comprise a truly honest, accurate presentation of the facts could make a Senate filibuster seem as fleeting as haiku. Thus, the goal of honesty has the potential or, even likelihood, to undermine efficacy. There are very few lawmakers, journalists or any others willing to read a tome heavy enough to double as a dumbbell. And in this day and age the so-called fifteen minutes of fame actually lasts about thirty seconds on television or about five minutes at a congressional hearing. As we've discussed, there is little room for nuance in the news or at most public hearings. To be effective, you have to be concise, lucid, memorable, and, if possible, colorful. These requirements are at odds with the requirements for honesty.

WHAT'S A SCIENTIST TO DO?

Schneider's advice, just as we have stressed throughout this book, is that when restrictions in time or space require simple, pithy statements, ensure your effectiveness with "accessible language and metaphors." In television and radio interviews and congressional hearings, therefore, Schneider is willing to simplify for the sake of effectiveness, but he supports his on-air statements with a "hierarchy of backup products": opinion editorials, articles for lay persons, websites, and books, each more detailed (and boring) than the previous.

Len Fisher, David Grinspoon, Jonathan Beckwith, and Stephen Schneider all have at least one important thing in common: they are all anxious to help the public understand scientific thinking and to inject rational, scientifically sound thinking into public policy. That may seem like a laudable, uncontroversial goal, but as their experiences show, this objective is difficult to attain and can be risky. Sometimes a strong constitution and personal sacrifice are required. Yet despite the hardships, these scientists—and many like them—say they are glad they charted the courses they did. They encourage others to follow in their paths, taking the advice of Mohandas Gandhi: "Be the change that you want to see in the world."[19] If we want a better Earth for future generations, many of the scientists in this book say their profession needs to help the public understand the value of its work—and of science in general. We hope this book has inspired you with at least one new way in which you can fulfill this civic duty.

Notes

INTRODUCTION

1 National Science Board, Science and Engineering Indicators 2004, vol. 1 (Arlington, VA: National Science Foundation, 2004), 7–16.

2 Ibid.

3 Ibid., 7–23.

4 Ibid.

5 Kevin Coyle, "Achieving Environmental Literacy in America" (Washington, D.C.: National Environmental Education & Training Foundation, 2005), ix.

6 Albert Einstein, Address at dedication of Museum of Science and Industry, February 11, 1936, Courtesy of the Einstein Archives Online, the Hebrew University of Jerusalem and the Einstein Papers Project, California Institute of Technology, 2003.

7 Greg Pearson and A. Thomas Young, eds., Technically Speaking (Washington, D.C.: National Academy Press, 2002), 1.

8 Coyle, "Achieving Environmental Literacy," 7.

9 Jacob Bronowski, Science and Human Values (New York: Harper & Row, 1956), 13.

CHAPTER 1: WE NEED TO TALK

1 Richard Hayes, unpublished Union of Concerned Scientists poll of scientist members.

2 Jim Hartz and Rick Chappell, Worlds Apart: How the Distance Between Science and Journalism Threatens America's Future (Nashville, TN: First Amendment Center, 1997).

3 Ibid., xii.

4 Ibid., 126–132.

5 Research study conducted by Market & Opinion Research International (MORI) for The Wellcome Trust, The Role of Scientists in Public Debate: Full Report, March 2000. Online at http://www.wellcome.ac.uk/assets/wtd003425.pdf.

6 Ibid., 17.

7 Some activists in Great Britain say that one reason why some scientists want better press is to counter widespread public opposition to the use of bovine growth hormone for increasing milk production, to genetically modified food, and to other new technologies. True or not, scientists on all sides of these issues will be able to use improved communications skills to make their views known to the public.

8 British Association for the Advancement of Science, "BA Media Fellowships." Online at http://www.the-ba.net/NR/rdonlyres/BF281954-2778-470F-981A-E05CEE956B10/0/MediaFellowsEvaluationReport.pdf.

9 James W. Tankard and Michael Ryan, "News Source Perceptions of Accuracy of Science News Coverage," Journalism Quarterly 51 (1974): 219.

10 Deborah M. Urycki and Stanley T. Wearden, "Science Communication Skills of Journalism Students," Newspaper Research Journal 19 (1998): 64.

11 One finding that substantiates the often-noted corrosive impact of broadcast television is that students who got more of their information from the print media instead of television wrote better articles.

12 Sheila Tobias, "Restructuring Supply, Restructuring Demand," Change 27 (1995): 22.

13 D. P. Phillips, E. J. Kanter, B. Becharezk, and P. L. Tastad, "Importance of the Lay Press in the Transmission of Medical Knowledge to the Scientific Community," New England Journal of Medicine 325 (1991): 1180–3.

14 Jennifer S. Haas, Celia P. Kaplan, Eric P. Gerstenberger, and Carla Kerlikowiske, "Changes in the Use of Postmenopausal Hormone Therapy After the Publication of Clinical Trial Results," Annals of Internal Medicine 140 (2004): 184–8.

15 A. Brzezinski, M. G. Vangel, Richard J. Wurtman, G. Norrie et al., "Effects of Exogenous Melatonin on Sleep: A Metanalysis," Sleep Medicine Reviews 9 (2005): 41–50.

16 National Science Board, Science and Engineering Indicators 2004, vol. 2 (Arlington, VA: National Science Foundation, 2004), A5–4.

17 National Science Board, Science and Engineering Indicators 2004, vol. 1, 4–30; and National Science Board, Science and Engineering Indicators 2004, vol. 2, A5–5.

18 Ibid.

19 National Science Foundation, "Merit Review Broader Impacts Crite-

rion: Representative Activities. Online at http://www.nsf.gov/pubs/
gpg/broaderimpacts.pdf.

CHAPTER 2: HOPE FOR THE BEST, PREPARE FOR THE WORST

1 Chris D. Thomas, et al. "Extinction Risk from Climate Change,"
  Nature 427 (2004): 145.

2 For example, James Gorman, "Scientists Predict Widespread Extinction by Global Warming," *New York Times*, 8 January 2004, late edition.

3 Jane Kay, "Dire Warming Warning for Earth's Species; 25% Could
  Vanish by 2050 as Planet Heats Up, Study Says," *San Francisco Chronicle*, 8 January 2004, final edition.

4 Tom Spears, "Warming Will Kill Off a Million Species in 50 Years:
  Report," Ottawa Citizen, 8 January 2004, final edition.

5 Letter to the Editor by Richard J. Ladle, Paul Jepson, Miguel B.
  Aráujo, and Richard J. Whittaker, "Dangers of Crying Wolf over Risk
  of Extinctions," *Nature* 428 (2004): 799.

6 Letter to the editor by Lee Hannah and Brad Phillips, "Extinction-
  Risk Coverage is Worth Inaccuracies," *Nature* 430 (2004): 141.

7 Steve Connor, "Scientists Find Prozac 'Link' to Brain Tumours," The
  Independent, 26 March 2002.

8 Ibid.

9 Adamantios Serafeim, et al., "5-Hydroxytryptamine Drives Apoptosis in Biopsylike Burkitt Lymphoma Cells: Reversal by Selective Serotonin Reuptake Inhibitors," Blood 99 (2002): 2545.

10 Martha C. Whiteman, I. J. Deary, A. J. Lee, and F.G.R. Fowkes, "Submissiveness and Protection From Coronary Heart Disease in the
   General Population: Edinburgh Artery Study," *Lancet* 350 (1997): 541.

11 David Fletcher, "Put Down That Rolling Pin Darling, It's Bad for Your
   Heart..." *Daily Telegraph* (UK), 21 August 1997, quoted in Ian J. Deary,
   Martha C. Whiteman, and F.G.R. Fowkes, "Medical Research and the
   Popular Media," *Lancet* 351 (1998): 1726.

12 Deary, Whiteman, and Fowkes, "Medical Research and the Popular
   Media."

13 Editorial, "Media Studies for Scientists," *Nature* 416 (2002): 461.

14 See, for example, S. Dunwoody and B. T. Scott, "Scientists as Mass
   Media Sources," Journalism Quarterly 59 (1982): 52–59; S. Dunwoody,
   "A Question of Accuracy," EEE Transactions on Professional Com-
   munication, PC25 4 (1982): 196–199; S. Dunwoody, "The Scientist as

a Source," in Scientists and Journalists: Reporting Science as News, ed. S. M. Friedman, S. Dunwoody, and C. L. Rogers (New York: The Free Press, 1986), 3–16; and J. W. Tankard and M. Ryan, "News Source Perceptions of Accuracy of Science Coverage," Journalism Quarterly 51 (1974): 219–225.

15 Vikki Entwistle, "Reporting Research in Medical Journals and Newspapers," British Medical Journal 310 (1995): 920.

16 Saira Bahl, Michelle Cotterchio, Nancy Kreiger, and Neil Klar, "Antidepressant Medication Use and Non-Hodgkin's Lymphoma Risk: No Association," American Journal of Epidemiology 160 (2004): 566.

17 Tineke Boddé, "Biologists and Journalists: A Look at Science Reporting," Bioscience 32 (1982): 173–175.

CHAPTER 3: WHY REPORTERS DO WHAT THEY DO

1 The Project for Excellence in Journalism, "The State of The News Media, 2004." Online at http://www.stateofthemedia.com/2004/narrative_networktv_contentanalysis.asp?cat=2&media=4

2 Ibid.

3 Scientists Institute for Public Information, "Now in 66 dailies: Newspaper Science Sections Spreading Nationwide," SIPIscope 14 (1986): 1–17.

4 Andrew Tyndall, unpublished data, www.tyndallreport.com; American Association of Advertising Agencies and the Association of National Advertisers, "Television Commercial Monitoring Report," 18. Internet use for news, however, has grown. According to a poll conducted by the Pew Center for the People and the Press, the number of people who consult the Internet at least three times a week has grown from 2 percent in 1996 to 25 percent in 2003. See The Pew Research Center for the People and the Press, "Public's News Habits Little Changed by September 11" (Washington, D.C.: The Pew Research Center, 2002). Online at http://people-press.org/reports/display.php3?PageID=613.

5 VCU Center for Public Policy, "Public Values Science But Concerned About Biotechnology," 2003. Online at http://64.233.167.104/u/vcu?q=cache:qmrVr3AUv1MJ:www.vcu.edu/lifesci/images2/PublicValues.pdf+public+values+science&hl=en&ie=UTF-8.

6 Stephen E. Bennet, Staci L. Rhine, and Richard S. Flickinger, "The Things They Care About: Change and Continuity in Americas' Attention to Different News Stories," The Harvard International Journal of Press/Politics 9 (2004): 84.

7   Pew Research Center, "Public's News Habits." Online at http://people-press.org/reports/display.php3?PageID=616 .Ibid.

8   Ibid.

9   Dan Rutz, "*Health Week*," Cable News Network (CNN), 2 June 1990.

10 Charles P. Alexander, "Medical Progress--Live! On CNN!" *Time* Magazine, 25 June 1990.

11 Judy Foreman, "Interest is Intense in an AIDS Heat Test," *Boston Globe*, 7 June 1990, city edition.

12 Paul Raeburn, "AIDS Patient Dies after Heat Treatment," Associated Press, 15 August 1990, A.M. cycle.

13 Summary of Findings: Site Visit Report: Clinical Use of Hyperthermia in AIDS, National Institute of Allergy and Infectious Disease, September 4, 1990.

14 State of Georgia, Composite State Board of Medical Examiners, Board Order 91-259, January 6, 1995. Online at http://www.medicalboard.state.ga.us/bdsearch/pdf/04-016011.pdf.

15 Paul Brooks "The House of Life: Rachel Carson at Work with Selections from Her Writings, Published and Unpublished" (Boston: Houghton Mifflin, 1972) quoted in Jon Fripp, Michael Fripp, and Deborah Fripp, Speaking of Science: Notable Quotes on Science, Engineering, and the Environment (Eagle Rock, VA: LLH Technology Publishing, 2000), 3.

16 John S. James, "Hyperthermia Report: Only One Patient," Aids Treatment News, 1 June 1990. Online at http://www.aids.org/atn/a-104-03.html.

17 Alexander, "Medical Progress."

18 The Project for Excellence in Journalism, "The State of The News Media, 2004." Online at http://www.stateofthemedia.com/2004/narrative_overview_ownership.asp?media=1.

19 The Wallace Stegner Initiative of the Institutes for Journalism and Natural Resources, Matching the Scenery: Journalism's Duty to the North American West, 2003, 62, 64.

20 The Project for Excellence in Journalism. "The State of The News Media, 2004." Online at http://www.stateofthemedia.com/2004/chartland.asp?id=221&ct=line&dir=&sort=&col1_box=1&col2_box=1#.

21 The Wallace Stegner Initiative of the Institutes for Journalism & Natural Resources, Matching the Scenery.

22 Ibid., 26.

23 Gary Schwitzer, "Merely Lights and Wires?" Minnesota Medical Association 86 (2003): 16–9.

24 Gary Schwitzer, "Ten Troublesome Trends in TV Health News," *British Medical Journal* 329 (2004): 1352.

25 Princeton Survey Research Associates, conducted for The Radio and Television News Directors Foundation, "Local Television News and Healthcare Survey: Report on the Findings Topline Results," 1995, 5.

26 Ibid., 23–4

27 Ibid., 5.

28 Statement by Daniel S. Goldin, NASA Administrator, 6 August 1996.

29 President Clinton Statement Regarding Mars Meteorite Discovery, 7 August 1996.

30 David McKay et al., "Search for Past Life on Mars: Possible Relic Biogenic Activity in Martian Meteorite ALH84001," Science 273 (1996): 924–930.

31 Kathy Sawyer, "NASA Releases Images of Mars Life Evidence; Space Agency Invites Further Inquiry by Others," Washington Post, 8 August 1996, final edition.

32 Robert C. Owen, "Martian Message--There Was Once Ancient Life Here," Christian Science Monitor, 8 August 1996.

33 Letter to the Editor by John F. Kerridge, "Mars Media Mayhem," Science 274 (1996): 161.

34 The Project for Excellence in Journalism. "The State of The News Media, 2004." Online at http://www.stateofthemedia.com/ 2004/narrative_localtv_contentanalysis.asp?cat=2&media=6

35 The Project for Excellence in Journalism. "The State of The News Media, 2005." Online at http://www.stateofthemedia.com 2005/narrative_networktv_contentanalysis.asp?cat=2&media=4

CHAPTER 4: DO YOU HEAR WHAT YOU'RE SAYING?

1 Mike King, "You've Asked, and Here Are Some Answers," The *Atlanta Journal-Constitution*, 26 July 2003, 13A.

2 Paula LaRocque, "A Simple Concept, Clarity," *Dallas Morning News*, 9 March 2003, 1F.

3 Sourcebook for Teaching Science website, http://www.csun.edu/~vceed002/health/docs/tv&health.html.

CHAPTER 5: MASTERING THE INTERVIEW

1 Francis Solomon, "Chatting with Reporters; Words on Words," Policy and Practice, 1 March 2005, 32.

2 Jim Pavia, "Guidelines for Talking to Reporters Do Exist," *Investment News*, 9 May 2005, 10.

3 Solomon, "Chatting with Reporters," 32.

4 Pavia, "Guidelines for Talking," 10.

5 Leonard Downie Jr, "Deep Throat: The Post and Watergate," 1 June 2005; 1:00 P.M. (Live Discussion), http://www.washingtonpost.com/wp-dyn/content/discussion/2005/06/01/DI2005060100769.html.

CHAPTER 6: A REPORTER'S MOST TRUSTED SOURCE: YOU

1 Andrew C. Revkin, "The Environment," in A Field Guide for Science Writers, 2nd ed., eds., Deborah Blum, Mary Knudson, Robin Maranty Henig (Oxford: Oxford University Press, 2005), 222.

2 www.stateofthenewsmedia.org and Federal Communications Commission.

3 Joseph B. Frazier, "Columbia River Spring Salmon at New Lows," Associated Press, 15 April 2005.

4 David Perlman, "Mass Extinction Comes Every 62 Million Years, UC Physicists Discover," *San Francisco Chronicle*, 10 March 2005, A7.

5 Denis Cuff, "Study Links Traffic, Student Ailments," *Contra Costa Times*, 22 February 2005, F4.

6 Guy Gugliotta, "Science Notebook," Washington Post, 9 May 2005, A7.

7 David Fleshler, "Scientists' Study Warns of Fate of Florida's Reefs," Knight Ridder Tribune News Service, *Bradenton Herald*, 20 March 2005, 15.

8 Patricia Smith, "N.C. Opposes Endangered Species Status for Oysters," Daily News, 16 April 2005.

9 Eric Hand, "Skeleton of Neandertal Reveals Bell-Shaped Being," *Seattle Times*, 13 March 2005, A6. Eric Hand writes for the *St. Louis Post-Dispatch*, but this article appeared in the *Seattle Times*.

10 Dawn Withers, "Scientists Say E. Coast Vulnerable to Tsunamis," *Chicago Tribune*, 9 February 2005, C6.

11 Mike Lee, "Reviving a River: A $626 Million, 50-Year Conservation Plan for the Colorado River Tries to Balance Needs of Native Habitat with People's Demand for Water," *San Diego Union-Tribune*, 27 March 2005, A1.

12 Tom Vogt, "Mount St. Helens: 25th Anniversary--Scientists Keep Tabs from a Distance; Cell Phones, Digital Images, GPS Technology Make Watching Volcano Today Vastly Different from 1980's Work," The Columbian, 15 May 2005, 30.

13 Mac Daniel, "Whale's Death Leaves Questions," *Boston Globe*, 14 December 2004, B5.

CHAPTER 7: CHOOSING THE RIGHT COMMUNICATION TOOLS

1 Bruce Dold, "Meet the Tribune Editorial Board; A Citizen Speaks Its Mind for 157 Years," *Chicago Tribune*, 26 December 2004, C6.

2 Art Coulson, "Oh, For Some Honest Discussion of the Issues," *St. Paul Pioneer Press*, 10 October 2004, 10B.

3 Dold, "Meet the Tribune Editorial Board," C6.

4 Pete Wasson, "Getting Your Thoughts into the Newspaper," Wausau Daily Herald, 9 April 2005, 6A.

5 Coulson, "Oh, For Some Honest Discussion," 10.

6 From archives of *San Jose Mercury News* website, by permission.

7 Dold, "Meet the Tribune Editorial Board," C6.

CHAPTER 8: THE SCIENTIST AS CELEBRITY AND ACTIVIST

1 Wook Kim, review of How to Dunk a Doughnut, by Len Fisher, Entertainment Weekly, 24 October 2003, iii.

2 Neal Lane, "The Civic Scientist and Science Policy," in Science and Technology Policy Yearbook (Washington, D.C.: American Association for the Advancement of Science, 1999), chap. 22.

3 Neal Lane, "A Perspective on American Science and Technology Policy--Storm Clouds on the Horizon" (speech presented at the University of Oklahoma, Norman, OK, 24 March 2005).

4 Len Fisher, "The Physics of Sex," Physics World 8 (1995): 76.

5 Len Fisher, "Physics Takes the Biscuit," *Nature* 397 (1999): 469.

6 For more about dunking biscuits and other examples of the science of the familiar, see Len Fisher, How to Dunk a Doughnut: The Science of Everyday Life (London: Orion Publishing Group, 2002); and Len Fisher, Weighing the Soul: The Evolution of Scientific Beliefs (London: Orion Publishing Group, 2004). For information about the Ig Nobel prize, see http://www.improbable.com/ig/ig-top.html.

7 Rae Goodell, The Visible Scientists (Boston: Little, Brown, 1977).

8 Carl Sagan, Cosmos (New York: Random House, 1980).

9 Lane, "A Perspective on American Science and Technology Policy."

10 David Grinspoon, Venus Revealed: A New Look Below the Clouds of Our Mysterious Twin Planet (Perseus Books Group, 1997).

11 Letter by Peter Spear, Dean, College of Arts and Sciences, University of Colorado at Boulder, "Tenure and Promotion Review for David Grinspoon," March 6, 1998.

12 David Grinspoon, *Lonely Planets*: The Natural Philosophy of Alien Life (New York: Ecco Press, 2003).

13 "David Grinspoon's World." Online at http://www.funkyscience.net/index.html.

14 Jon Beckwith, Making Genes, Making Waves: A Social Activist in Science (Cambridge, MA: Harvard University Press, 2002).

15 Ibid., 55.

16 P. Jacobs, et al., "Aggressive Behavior, Subnormality, and the XYY Male," *Nature* 208 (1965): 1351–1352.

17 Hongping Tian, Dana Boyd, and Jon Beckwith, "A Mutant Hunt for Defects in Membrane Protein Assembly Yields Mutations Affecting the Bacterial Signal Recognition Particle and Sec Machinery," PNAS 97 (2000): 4730–4735.

18 Sharon M. Friedman, Sharon Dunwoody, and Carol L. Rogers, eds., Communicating Uncertainty: Media Coverage of New and Controversial Science (Mahwah, N.J.: Lawrence Erlbaum Associates, 1999), 91.

19 Jon Fripp, Michael Fripp, and Deborah Fripp, Speaking of Science: Notable Quotes on Science, Engineering, and the Environment (Eagle Rock, VA: LLH Technology Publishing, 2000), 200.

Resources

GROUPS WORKING ON SCIENCE
AND THE MEDIA

Aldo Leopold Leadership Program
Stanford Institute for the Environment
Encina Modular C, 429 Arguello Way
Stanford, CA 94305
(650) 725-0651
info@leopoldleadership.org
www.leopoldleadership.org

AAAS Mass Media Science &
Engineering Fellows Program,
American Association for the
Advancement of Science
1200 New York Ave., NW
Washington, D.C. 20005
(202) 326-6441
www.aaas.org/programs/education/
MassMedia/program.shtml

National Association of Science
Writers
P.O. Box 890
Hedgesville, WV 25427
(304) 754-5077
www.nasw.org

Society of Environmental Journalists
P.O. Box 2492
Jenkintown, PA 19046
(215) 884-8174
http://www.sej.org

Union of Concerned Scientists
2 Brattle Square
Cambridge, MA 02238
(617) 547-5552

ucs@ucsusa.org
www.ucsusa.org

MEDIA RELATIONS CONSULTANTS

Caplan Communications (specializes
in satellite TV tours and radio tours)
12530 Parklawn Drive, Suite 250
Rockville, MD 20852
(301) 770-0550
ccinfo@caplancommunications.com
www.caplancommunications.com/

Morrow Cater, Cater Communications (message development, communications strategy, media releases)
179 Reservoir Road
San Rafael, CA 94901
(415) 453-0430
morry@catercommunications.com
www.catercommunications.com

PRESS RELEASE
DISTRIBUTION SERVICES

EurekAlert!/AAAS
1200 New York Ave., NW
Washington, D.C. 20005
(202) 326-6716
webmaster@eurekalert.org
www.eurekalert.org

Newswise
930 Turner Mountain Road
Charlottesville, VA 22903
(434) 296-9417
info@newswise.com
www.newswise.com

PR Newswire
810 7th Ave., 35th fl.
New York, NY 10019
(800) 832-5522
information@prnewswire.com
www.prnewswire.com

U.S. Newswire
National Press Building, Suite 1230
Washington, D.C. 20045
(800) 544-8995
info@usnewswire.com
www.usnewswire.com

MEDIA TRAINING FOR SCIENTISTS

Aldo Leopold Leadership Program
Stanford Institute for
the Environment
Encina Modular C, 429 Arguello Way
Stanford, CA 94305
(650) 725-0651
info@leopoldleadership.org
www.leopoldleadership.org

AAAS Mass Media Science &
Engineering Fellows Program
American Association for the
Advancement of Science
1200 New York Ave., NW
Washington, D.C. 20005
(202) 326-6441
www.aaas.org/programs/education/
MassMedia/program.shtml

Christine K. Jahnke, President
Positive Communications
5726 MacArthur Blvd., NW
Washington, D.C. 20016
(202) 393-0764
poscom@erols.com
www.poscom.com

Richard Hayes, Media Director
Union of Concerned Scientists

1701 H Street, Suite 600
Washington, D.C. 20006
202-223-6133
www.ucsusa.org
rhayes@ucsusa.org

SCIENCE FELLOWSHIPS AND
TRAINING PROGRAMS FOR
JOURNALISTS

Ted Scripps Fellowship (Academic-
year fellowships for environmental
journalists to improve their understanding of environmental topics)
University of Colorado
1511 University Ave., 478 UCB
Boulder, CO 80309-0478
(303) 492-0459
www.colorado.edu/journalism/cej/

Science Literacy Project
(One-week science journalism training for mid-career radio producers
and reporters)
SoundVision Productions
2991 Shattuck Ave., Suite 304
Berkeley, California 94705-1872
(510) 486-1185
www.scienceliteracyproject.com

Knight Science Journalism Fellowships (Academic-year fellowships to
improve understanding of science
and one-week "boot camps" on specific areas of science reporting)
77 Massachusetts Ave., MIT E32-300
Cambridge MA 02139
(617) 253-3442
http://web.mit.edu/knight-science/
fellowships.html

Institute for Journalism and
Natural Resources
(One-week expeditions for reporters,
photojournalists, writers, editors

and producers to learn about natural resource and land use issues)
121 Hickory Street, Suite 2
Missoula, Montana, 59801
(406) 273-4626
http://www.ijnr.org/programs/expeditions.htm

Ocean Science Journalism Fellowship (One-week workshop with possible extensions for independent study on ocean sciences and engineering)
Woods Hole Oceanographic Institution
Fenno House MS#40
Woods Hole, MA 02543
(508) 289-2270
www.whoi.edu/home/news/media_jfellowship.html

MBL Science Journalism Program (One-week fellowships with possible extensions for independent study in environmental and biomedical research)
Marine Biological Laboratory
7 MBL Street
Woods Hole, MA 02543-1015
(508) 289-7423
http://www.mbl.edu/inside/what/news/sci_journal/index.html

VIDEO AND AUDIO NEWS RELEASES

Home Front Communications
1620 I Street, NW #520
Washington, D.C. 20006
(202) 544-8400

Medialink
708 Third Ave., New York, NY 10017
(800) 843-0677
info@medialink.com
www.medialink.com

NSF GRANTS FOR COMMUNICATING SCIENCE

Communicating Research to Public Audiences
National Science Foundation
4201 Wilson Blvd.
Arlington, VA 22230
(703) 292-5127
http://www.nsf.gov/funding/pgm_summ.jsp?pims_id=5362&from=fund

TOP 100 U.S. NEWSPAPERS

Akron Beacon Journal
44 East Exchange Street
P.O. Box 640, Akron, OH 44309-0640
(330) 996-3000
www.ohio.com

Albuquerque Journal
7777 Jefferson Street
Albuquerque, NM 87109
(505) 823-3800
www.abqjournal.com

Arizona Republic
200 East Van Buren Street
Phoenix, AZ 85004
(602) 444-8000
www.azcentral.com/arizonarepublic/

Arkansas Democrat-Gazette
P.O. Box 2221
Little Rock, AR 72203-2221
(501) 378-3400
www.ardemgaz.com

Arlington Heights Daily Herald
P.O. Box 280
Arlington Heights, IL 60006-0280
(847) 427-4300
www.dailyherald.com

Asbury Park Press
3601 Highway 66
P.O. Box 1550, Neptune, NJ 07754
(732) 922-6000
www.app.com

Atlanta Journal-Constitution
P.O. Box 4689, Atlanta, GA 30302
(404) 222-2025
www.ajc.com

Austin American Statesman
P.O. Box 670
Austin, TX 78767
(512) 353-5022
www.statesman.com

Baltimore Sun
P.O. Box 1377
Baltimore, MD 21278
(410) 332-6000
www.baltimoresun.com

Bergen County Record
150 River Street
Hackensack, NJ 07601
(201) 646-4100
www.njmg.com

Birmingham News
P.O. Box 2553
Birmingham, AL 35202-2553
(205) 325-2222
www.bhamnews.com

Boston Globe
135 Morrissey Blvd.
Boston, MA 02107
(617) 929-2000
www.boston.com

Boston Herald
P.O. Box 55843
Boston, MA 02205
(617) 426-3000
www.bostonherald.com

The Buffalo News
P.O. Box 100, Buffalo, NY 14240
(716) 849-4444
www.buffalonews.com

Charlotte Observer
600 South Tryon Street
Charlotte, NC 28232
(704) 358-5000
www.charlotte.com

Chicago Sun-*Times*
350 N. Orleans
Chicago, IL 60654
(312) 321-3000
www.suntimes.com

Chicago Tribune
435 N. Michigan Ave.
Chicago, IL 60611
(312) 222-3232
www.chicagotribune.com

Cincinnati Enquirer
312 Elm Street
Cincinnati, OH 45202
(513) 721-2700
http://news.enquirer.com/

Columbus Dispatch
34 S. 3rd Street
Columbus, OH 43215
(614) 461-5000
www.dispatch.com

The Commercial Appeal
495 Union Ave.
Memphis, TN 38103
(901) 529-2345
www.commercialappeal.com

Contra Costa Times
2640 Shadelands Drive
Walnut Creek, CA 94598
(925) 935-2525
www.contracostatimes.com

The Courier-Journal
525 W. Broadway
P.O. Box 740031
Louisville, KY 40201-7431
(502) 582-4011
www.courier-journal.com

The *Dallas Morning News*
P.O. Box 655237
Dallas, TX 75265-5237
(214) 977-8359
www.dallasnews.com

Dayton Daily News
2150 North Gettysburg Ave.
Dayton, OH 45406
(937) 263-0172
www.daytondailynews.com

Denver Post
1560 Broadway
Denver, CO 80202-6000
(303) 820-1557
www.denverpost.com

Des Moines Register
P.O. Box 957
Des Moines, IA 50304-0957
(515) 284-8000
www.desmoinesregister.com

Detroit Free Press
600 W. Fort
Detroit, MI 48226
(313) 222-2441
www.freep.com

The Detroit News
615 W. Lafayette
Detroit, MI 48226
(313) 222-2300
www.detnews.com

Express News
P.O. Box 2171
San Antonio, TX 78297

(210) 250-3000
www.mysanantonio.com

Florida *Times* Union
1 Riverside Ave.
Jacksonville, FL 32202
(904) 359-4111
www.jacksonville.com/

Fort Worth Star Telegram
400 West 7th
Fort Worth, TX 76102
(817) 390-7400
www.dfw.com

Fresno Bee
1626 E Street
Fresno, CA 93786-0001
(559) 441-6111
www.fresnobee.com

Grand Rapids Press
155 Michigan Street, NW
Grand Rapids, MI 49503
(616) 222-5400
www.mlive.com/grpress

The Hartford Courant
285 Broad St
Hartford, CT 06115-3785
(860) 241-6200
www.courant.com

The Honolulu Advertiser
605 Kapiolani Blvd.
Honolulu, HI 96813
P.O. Box 3110
Honolulu, HI 96802
(808) 525–8090
www.honoluluadvertiser.com

Houston Chronicle
P.O. Box 4260
Houston, TX 77210-4260
(713) 220-7171
www.chron.com

Indianapolis Star
P.O. Box 145
Indianapolis, IN 46206
(317) 444-4000
www.indystar.com

The Journal News
1 Gannett Drive
White Plains, NY 10604
(914) 694-9300
www.thejournalnews.com

Kansas City Star
1729 Grand Blvd.
Kansas City, MO 64108
(816) 234-4636
www.kansascity.com/mld/
kansascity/

The Knoxville News Sentinel
2332 News Sentinel Drive
Knoxville, TN 37921-5761
(865) 523-3131
www.knoxnews.com/kns/news/

LA Investor's Business Daily
12655 Beatrice Street
Los Angeles, CA 90066-7300
(310) 448-6700
www.investors.com

Las Vegas Review Journal
P.O. Box 70
Las Vegas, NV 89125
(702) 383-0211
www.reviewjournal.com

Lexington Herald-Leader
100 Midland Ave.
Lexington, KY 40508
(859) 231-3100
www.kentucky.com

Los Angeles Daily News
P.O. Box 4200

Woodland Hills, CA 91365
(818) 713-3000
www.dailynews.com

Los Angeles La Opinion
411 W 5th Street
Los Angeles, CA 90013
(213) 622-8332
www.laopinion.com

Los Angeles Times
202 W. 1st Street
Los Angeles, CA 90012
(213) 237-5000
www.latimes.com

Miami Herald
One Herald Plaza
Miami, FL 33132
(305) 350-2111
www.miami.com

Milwaukee Journal Sentinel
P.O. Box 661
Milwaukee, WI 53201
(414) 224-2000
www.jsonline.com

Minneapolis Star Tribune
425 Portland Ave. S.
Minneapolis, MN 55488
(612) 673-4000
www.startribune.com

The Morning Call
101 North 6th Street
Allentown, PA 18101
(610) 820-6500
www.mcall.com

News and Observer
P.O. Box 191
Raleigh, NC 27602
(919) 829-4700
www.newsobserver.com

Newsday
235 Pinelawn Drive
Melville, NY 11747-4226
(516) 843-3403
www.newsday.com

The News Journal
P.O. Box 15505, Wilmington, DE 19850
(302) 324-2500
www.delawareonline.com

The News Tribune
1950 South State Street
Tacoma, WA 98405
(253) 597-8742
www.thenewstribune.com

New York Daily News
450 W. 33 Street
New York, NY 10001
(212) 210-2100
www.nydailynews.com

The New York Post
1211 Avenue of the Americas
New York, NY 10036-8790
(212) 930-8000
www.nypost.com

The *New York Times*
229 West 43d Street
New York, NY 10036
(646) 223-3260
www.nytimes.com

The Oklahoman
9000 Broadway Extension
Oklahoma City, OK 73114
(405) 475-3311
www.newsok.com

Omaha-World Herald
World-Herald Square
Omaha, NE 68102
(402) 444-1000
www.omaha.com

Orange County Register
625 N. Grand Ave.
Santa Ana, CA 92701
(877) 469-7344
www.ocregister.com

The *Oregonian*
1320 SW Broadway
Portland, OR 97201
(503) 221-8160
www.oregonian.com

Orlando Sentinel
633 N. Orange Ave., Orlando, FL 32801
(407) 420-5000
www.orlandosentinel.com

The Palm Beach Post
P.O. Box 24700
West Palm Beach, FL 33416
(561) 820-4400
www.palmbeachpost.com

Philadelphia Daily News
400 North Broad Street
Philadelphia, PA 19130
(215) 854-5900
www.philly.com

The *Philadelphia Inquirer*
P.O. Box 8263, Philadelphia, PA 19101
(215) 854-2000
www.philly.com

Pittsburgh Post-Gazette
34 Blvd of the Allies
Pittsburgh, PA 15222
(412) 263-1100
www.post-gazette.com

Pittsburgh Tribune-Review
D.L. Clark Bldg.
503 Martindale Street, 3rd fl.
Pittsburgh, PA 15212
(412) 321-6460
www.pittsburghlive.com/x/

The Plain Dealer
1801 Superior Ave E.
Cleveland, OH 44114
(216) 999-5000
www.plaindealer.com

The Post-Standard
Clinton Square
P.O. Box 4915
Syracuse, NY 13221-4915
(315) 470-0011
www.post-standard.com

Press-Enterprise
3512 14th Street
Riverside, CA 92501
(909) 684-1200
www.pe.com

Providence Journal-Bulletin
75 Fountain Street
Providence, RI 02902
(401) 277-7000
www.projo.com

Richmond *Times* Dispatch
300 E. Franklin Street
Richmond, VA 23219
(804) 649-6000
www.timesdispatch.com

Rochester Democrat & Chronicle
55 Exchange Blvd.
Rochester, NY 14614-2001
(585) 258-2280
www.democratanD.C.hronicle.com

Rocky Mountain News
100 Gene Amole Way
Denver, CO 80204
(303) 892-5000
www.rockymountainnews.com

Sacramento Bee
2100 Q Street
Sacramento, CA 95816

(916) 321-1000
www.sacbee.com

Salt Lake Tribune
90 S. 400 West, Suite 700
Salt Lake City, UT 84101
(801) 257-8742
www.sltrib.com

San Diego Union-Tribune
P.O. Box 120191
San Diego, CA 92112-0191
(619) 299-3131
www.signonsandiego.com

San Francisco Chronicle
901 Mission Street
San Francisco, CA 94103
(415) 777-1111
www.sfgate.com/chronicle/

San Jose Mercury News
750 Ridder Park Dr.
San Jose, CA 95190-0001
(408) 920-5000
www.mercurynews.com

Sarasota Herald Tribune
801 S. Tamiami Trail
Sarasota, FL 34236
(941) 953-7755
www.newscoast.com

Seattle Post Intelligencer
P.O. Box 1909
Seattle, WA 98111-1909
(206) 448-8000
http://seattlepi.nwsource.com/

Seattle Times
P.O. Box 70
Seattle, WA 98111
206-464-2111
http://seattletimes.nwsource.com

South Florida Sun-Sentinel
200 E. Las Olas Blvd.
Fort Lauderdale, FL 33301
(954) 356-4000
www.sun-sentinel.com

Spokane Spokesman Review
P.O. Box 2160
Spokane, WA 99210
(509) 459-5000
www.spokesmanreview.com

The Star-Ledger
1 Star Ledger Plaza
Newark, NJ 07102-1200
(973) 392-4141
www.starledger.com

The State
P.O. Box 1333
Columbia, SC 29202-1333
(803) 771-6161
www.thestate.com

St. Louis Post-Dispatch
900 N. Tucker Blvd.
St. Louis, MO 63101
(314) 340-8000
www.stltoday.com

St. Paul Pioneer Press
345 Cedar Street, St. Paul, MN 55101
(651) 222-1111
www.twincities.com

St. Petersburg Times
490 First Ave. S.
St. Petersburg, FL 33701
(727) 893-8111
http://sptimes.com/

Tampa Tribune
200 S. Parker Street
Tampa, FL 33606
(863) 683-6531
http://tampatrib.com

Telegram & Gazette
20 Franklin Street
Box 15012, Worcester, MA 01615-0012
(508) 793-9100
www.telegram.com

The Tennessean
1100 Broadway, Nashville, TN 37203
(615) 259-8300
http://tennessean.com

The *Times*-Picayune
3800 Howard Ave.
New Orleans, LA 70125-1429
(504) 826-3300
www.timespicayune.com

Toledo Blade
541 N. Superior Street
Toledo, OH 43660
(419) 724-6000
www.toledoblade.com

Tulsa World
315 S. Boulder Ave.
Tulsa, OK 74103-3401
(918) 583-2161
www.tulsaworld.com

USA Today
7950 Jones Branch Drive
McLean, VA 22108-0605
(703) 854-7121
www.usatoday.com

Virginian Pilot
150 West Brambleton Ave.
Norfolk, VA 23510
(757) 446-2989
http://welcome.hamptonroads.com/

Wall Street Journal
200 Liberty Street
New York, NY 10281
(212) 416-2000
www.wsj.com

Washington Post
1150 15th Street, NW
Washington, D.C. 20071-0001
(202) 334-6173
www.washingtonpost.com

Washington *Times*
3600 New York Ave. NE
Washington, D.C. 20002-1996
(202) 636-3000
www.washingtontimes.com

MAJOR MAGAZINES

National Geographic
1145 17th Street, NW
Washington, D.C. 20036
(202) 857-7000
www.nationalgeographic.com

Nature
968 National Press Building
529 14th Street NW
Washington, D.C. 20045-1938
(202) 737 2355
www.nature.com

New Scientist
Lacon House
84 Theobald's Road
London WC1X8NS
44 (0)20 7611 1200
www.newscientist.com

Newsweek
251 West 57th Street
New York, NY 10019
(212) 445-4000
www.newsweek.msnbc.com

Science
1200 New York Ave., NW
Washington, D.C. 20005
(202) 326-6500
www.sciencemag.org

Science News
1719 N Street, NW
Washington, D.C. 20036
(202) 785-2255
www.sciencenews.org

Scientific American
415 Madison Ave., New York, NY 10017
(212) 754-0550
www.sciam.com

Smithsonian Magazine
P.O. Box 37012 MRC 951
Washington, D.C. 20013
(202) 275-2000
www.smithsonianmag.si.edu

Time
Time & Life Building
Rockefeller Center
New York, NY 10020
(212) 522-1212
www.time.com

U.S. News and World Report
1050 Thomas Jefferson Street, NW
Washington, D.C. 20007
(202) 955-2000
www.usnews.com

MAJOR TELEVISION NEWS
PROGRAMS AND CABLE NEWS
NETWORKS

ABC World News Tonight
47 West 66th Street, 2nd fl.
New York, NY 10023
(212) 456-4040
www.abcnews.com

CBS Evening News
524 West 57th Street
New York, NY 10019
(212) 975-4321
www.cbs.com

CNN
One CNN Center, Atlanta, GA 30303
(404) 827-1500
www.cnn.com

Fox News Channel
1211 Avenue of the Americas
New York, NY 10036
(212) 301-3000
www.foxnews.com

MSNBC
One MSNBC Plaza
Secaucus, NJ 07094
(201) 583-5000
www.msnbc.com

NBC Nightly News
30 Rockefeller Plaza
New York, NY 10112
(212) 664-4971
www.nightlynews.msnbc.com

The NewsHour (PBS)
3620 27th Street South
Arlington, VA 22206
(703) 998-2150
www.pbs.org/newshour

The Science Channel
One Discovery Place
Silver Spring, MD 20910
(240) 662-2000
http://science.discovery.com

MAJOR RADIO NEWS NETWORKS

ABC News Radio
125 West End Ave., 6th fl.
New York, NY 10023
(212) 456-5100
www.abcradio.com

CBS Radio Network
524 West 57th Street
New York, NY 10019

(212) 975-3615
www.cbsradio.com

Hispanic Radio Network
1101 Pennsylvania Ave. NW, 6th fl.
Washington, D.C. 20004
(202) 637-8800
www.hrn.org

Living on Earth
(airs on public radio stations)
20 Holland Street, Suite 408
Somerville, MA 02144
(800) 218-9988
www.loe.org

National Public Radio
635 Massachusetts Ave., NW
Washington, D.C. 20001
(202) 513-2785 (science desk)
www.npr.org

NBC Radio Network
2020 M Street, NW
Washington, D.C. 20036
(202) 457-7990
www.westwoodone.com

Public Radio International
100 North Sixth Street, Suite 900-A
Minneapolis, MN 55403
(612) 338-5000
www.pri.org

Science Friday (NPR)
55 W. 45th Street, 4th fl.
New York, NY 10036
(212) 302-0804
www.sciencefriday.com

The Infinite Mind
(airs on public radio stations)
One Broadway, 14th fl.
Cambridge, MA 02142
(617) 682-3700
http://lcmedia.com/mindprgm.htm

USA Radio Network
2290 Springlake Road, Suite 107
Dallas, Texas 75234
(972) 692-1330
www.usaradio.com

Westwood One
2020 M Street, NW
Washington, D.C. 20036
(202) 457-7990
www.westwoodone.com

Voice of America
330 Independence Ave., SW
Washington, D.C. 20237
(202) 203-4302
www.voa.gov

Index

ABOUT THE AUTHORS

RICHARD HAYES is media director of the Union of Concerned Scientists, where for more than a dozen years he has helped scientists across the country become more effective with the press. Hayes was previously a reporter for a bipartisan caucus in the U.S. Congress, where he also coordinated a task force on climate science.

DANIEL GROSSMAN is an award-winning science journalist and former reporter for National Public Radio's show on the environment, *Living on Earth*. He has written for publications such as the *New York Times, Rolling Stone,* and *Scientific American*. He has also taught science journalism at the Boston University School of Journalism.